The Unobservable Universe

The Unobservable Universe

A Paradox-Free Framework for Understanding the Universe

Scott M. Tyson

GALAXiA WAY

Galaxia Way
Albuquerque, New Mexico

Published by: Galaxia Way
                     9312 Galaxia Way NE
                     Albuquerque, NM 87111
                     www.galaxiaway.com

Cover design by RHgraphx.com
Interior design by MediaNeighbours.com
Typeset in Garamond, Franklin Gothic, and Gill Sans

This paperback edition was produced in April 2011.
The Unobservable Universe is also available in eBook and audio book editions.
For more information, please visit www.theunobservableuniverse.com.

Library of Congress Control Number: 2011928217
ISBN-13: 978-0-9832438-0-9
ISBN-10: 0983243808

Publisher's Cataloging-in-Publication Data
Tyson, Scott M.
The Unobservable Universe : A Paradox-Free Framework for Understanding the Universe / Scott M. Tyson.
p. cm.
Includes bibliographical references and index.
1. Science—Cosmogony—Cosmology. 2. Cosmology—Science—Philosophy.
3. Philosophy—Cosmology. I. Title

Manufactured in the United States of America

10 09 08 07 06 05 04 03 02 01

I dedicate this book to my lovely and loving wife, Janna, who supported me tirelessly through this book project.

I also dedicate this book to the memory of Mr. Charles M. Ferri, my high school AP Chemistry teacher, whose whacky teaching style motivated my life's work.

Table of Contents

Section III. Implications, Confirmations, and Demonstrations: The End of Paradox

Figures and Photographs

Foreword

Virtually everything you've ever learned about anything is wrong!

Thus begins author Scott M. Tyson's enthralling journey through the far reaches of time and space—the final frontier—as he brings rocket science down to Earth and demystifies the universe for all of us.

For centuries humans have marveled at the stars and the seemingly infinite expanse of blackness that surrounds their home planet. In an attempt to understand the complexities of the universe, they created mythological creatures and spirits that carried the planets across the stars and gave names to collections of stars.

Tyson has devoted thirty-four years of his illustrious career as a trailblazing physicist and innovative scientist to the birth of this book. Indeed, such an endeavor brings to mind the birthing of the very stars themselves.

It all starts with an enormous sphere of hydrogen gas that becomes denser with the passage of time. As the compression builds, the temperature at the core rises even more. Eventually, the temperature approaches 10-million degrees kelvin, setting off a process similar to a hydrogen bomb explosion, or nuclear fusion. The ball of gas burns and a new star is born—a process that requires millions of years.

So, perhaps, thirty-four years represents a mere speck in the cosmic perspective of time for the development of Tyson's theories presented herein, which we believe will have a far-reaching impact similar to the evolutionary process of the stars described above.

Before humans developed the technology to jet-propel objects into space, they sought out the universe to assist in learning more about the Earth. This was the dawn of science, as described by the early writers of Western thought.

As the world became more understandable, humans turned once again toward the stars—this time in an attempt to understand with greater depth where Earthlings fell in relation to the cosmos.

A new age is dawning in man's capacity to understand the universe. And many theories abound concerning the origins of it. Tyson stands on the precipice of this new wave of human consciousness. Here, at last, in *The Unobservable Universe*, is an audacious blueprint for a new cosmogony that changes virtually everything.

It has been said that truth enjoys only a brief moment of victory between the time it is condemned as paradox and the time it is embraced as trivial. In *The Unobservable Universe*, Tyson addresses some of the questions facing cosmologists, astrophysicists, and quantum physicists, first by addressing the existence of paradoxes between the various divisions of science. Embracing the fundamental truth that greater understanding can reduce the existence of paradoxes in human knowledge, Tyson makes bold strides in exposing the holes in human knowledge by introducing new insights—innovative ways to think about what is accepted as truth.

Tyson's research starts out with a simple hypothesis: In order to reach some level of understanding, we must first acknowledge that not everything we have been taught over the years is the truth. The critical mind must continually assess incoming information, test it, and employ systems that enable better understanding. As he writes, "Truth seekers are particularly and keenly aware that perfection is unachievable; therefore, the prevailing body of knowledge contains inaccurate and incomplete content, since they themselves seek to make contributions to advance the state of knowledge and understanding."

Tyson takes us through a series of thought experiments meant to expose our vulnerability. How do we come to understand the limitations of our ability to perceive? If we can accept that some things are visible with the human eye, can we also accept that there are some things that are invisible to the human eye? If a tree falls in the middle of the woods and no one is around to hear it, does it make a sound? These are some of the questions that Tyson asks his readers, with a particular emphasis on the recognition of human limitations.

Our perceptions, Tyson argues, make up the basis of human culture. The science community—itself a culture, as he defines the term—is not immune from this distinction. Cultures have their own unique languages and ways of addressing certain phenomena, but they are all based on the ability to perceive. Is this also true of those who "talk in maths," pulling from a character in the song "Karma Police"? "Like any other culture," Tyson says while addressing the so-called "hard" scientists, "what is accepted and known is comfortable, the 'known devil,' while new thoughts, concepts, and approaches represent new and somewhat scary things."

Observation and perception are even heavily dependent on the language used to describe phenomena. Throughout the book we have included some notes that are intended to engage the reader in further questioning of the use of language. Many of the concepts that we accept as truth today were derived from pioneers who wrote, talked, and thought in a language different from American English. Anyone who has taken a foreign language can understand the complexities that can arise in interpreting a work that was written in another language.

In one section, Tyson discusses early understandings of void space by Greek philosophers—something we now understand to be a critical component of understanding the nature of the universe. He quotes Parmenides as saying: "Being is, while nothing is not"—a widely accepted translation of this axiom. He says Parmenides "felt that the void is the ultimate non-existence." A literal translation, however, brings out new insights: "To be, it is, as to not be is not." Further discussion of this can be found in the References section at the end of the book.

The point here is that with this basic technique, the true genius of Tyson's work shines forth. By placing people—observers and perceivers—in proper context and by coming closer to understanding human limitation, we have a better capacity to zero in on the paradoxes that have held us with a tight grip for time immemorial. With *The Unobservable Universe*, he intends to start a discussion by presenting a way of getting beyond the paradoxes found in widely accepted theories about our universe.

Finally, a discussion about the nature of the universe would not be complete without engaging the minds of great astrophysicists, quantum physicists, astronomers, and mathematicians—both living and dead. As if holding a conversation with Einstein himself, he states, quite candidly, that Einstein's "Theory of Relativity is quite specific in describing the way in which electromagnetic force travels through free space, but his theory doesn't define the internal nature of free space."

So, Einstein, now that we have an understanding of how light travels through space—thanks to you—and we understand the nature of gravity—thanks to you—what is the nature of space, anyway? A few theories have certainly emerged to explain what Einstein could not. Einstein himself built his theories on relativity from the scholarship that had preceded him. In like manner, with carefully crafted precision, Tyson pulls from his knowledge of the scholarship of others to address the paradoxical elements that still exist.

Perhaps the underlying brilliance of *The Unobservable Universe* lies in its remarkable literary style that speaks to all levels of scientific knowledge, including the simplest. Instead of the usual dry voice of most scientific tomes, Tyson threads his writing with uproarious wit, molecules of wisdom and raucous humor while expounding his case with unusual aplomb. Furthermore, he presents complex data with a rock-solid command of his subject, without ever sounding presumptuous or pedantic, to create an extraordinary cosmology treatise.

The beauty of this book is that Tyson never loses touch with his readers and even prompts them to "take a break" in order to digest some abstract or detailed concepts. Imagine that! He anticipates their questions and leaves no stone unturned to ensure they comprehend every point.

The greatest books that changed the world were those which stirred controversy. We truly believe that Tyson's work fulfills this definition and will undoubtedly open many minds in the process. Blending popular culture and scientific study into a melting pot that entertains as much as it educates, Tyson's artful style will delight cosmology buffs in a refreshing read that is fun and informative while captivating advanced cosmologists with compelling theories that will ignite debate for decades.

According to Tyson, the key to developing a new perspective from which to fully understand our universe is, at its very core, a challenge that involves deep cultural change. Science can then rapidly move beyond its current efforts aimed at trying to pry tidbits of understanding from our universe.

This groundbreaking book holds the potential to directly impact global warming, climate change, and other dangers associated with today's energy production. Through Tyson's perspectives, this book holds the potential to impact the lives of every person on the planet by providing the means to tap into the energy of the cosmos, which fuels all physical phenomena of our universe.

Tyson trailblazes a path that motivates humanity permanently away from fossil-fuel and nuclear-based energy production and toward a nearly unlimited and entirely benign energy paradigm that fosters the health and security of the planet.

As a result, science and humanity can instead embark upon a new age in which science harnesses the energy of the cosmos to provide a safer world from which humanity can more confidently and competently provide for the needs of all the Earth's inhabitants and continue its exploration outward to new worlds.

If for no other reason, Tyson prompts the reader to question the very foundations of everything he or she has ever learned. This is the only way that we, as a species, can accept ourselves for who we truly are as we strive to enhance our relationship with others. In that, *The Unobservable Universe* represents a timely gift for a world poised on the brink of a dramatic shift. It offers humanity an entirely new perspective from which to consider its oneness with the surrounding world and cross the threshold into a brave future.

We have never seen a book like this one—a bold, brilliant and exhilarating read that challenges scientific doctrine while offering a provocative proposal for the future of cosmology. We believe it will go down in the history of the world as a benchmark that forever changed it.

Let the debates begin!

Catherine J. Rourke
Editor and Award-Winning Journalist

Thomas M. Hill
Editor

Preface

Do you remember when you were seventeen? I do. And, as much as I would like to think that my memories were entirely typical, I seriously doubt it based on comments I've received from others.

It was 1976 and I was a junior at Hicksville High School on Long Island. (Yes, there's a place called Hicksville, New York; it's located midway between the north and south shores of Long Island, about thirty miles or so outside of Manhattan. No one would ever believe that there was actually a place called Hicksville; I was forever having to pull out my library card to prove it!) There are lots of things I recall about that year.

I recall participating on the track and field team and the cross country team. I recall working on the yearbook committee. I recall participating in the math and Latin clubs. I remember getting my driver's license the year before and spending lots of time exploring Long Island in a jalopy of a car that my father handed down to me and for which I was deeply grateful. I remember the girls I dated and the friends with whom I spent my time. I remember the music and the movies of that time period. I remember the year-long celebration of the nation's Bicentennial. I remember learning to write programs in BASIC on my school's fancy new Digital Equipment Corporation's Edusystem 20-based mainframe computer, which ran off input in the form of punched tape. I remember my parents and my teachers, both of whom continuously bugged me to strive for my best. I remember things I'm just not willing to share with readers of this book out of my senses of modesty and propriety. But, none of these things is the one thing I remember most.

What I remember most, by far, was a single period of a single high school science class. I was sitting in Mr. Ferri's class, and he had recently begun to cover a section on quantum mechanics. I was enthralled. Nothing, I mean nothing, ever engrossed me the way that learning about quantum mechanics did. The fact that

Mr. Ferri was the kind of teacher for whom it was simply implausible to teach in a dry, traditional, dull way probably helped, too. He was quite a character and he knew precisely how to keep the attention of his students. Mr. Ferri wore a gym teacher's whistle on a rope around his neck and he wasn't afraid to use it as he positively gyrated and pranced around the classroom with maniacal exuberance in his white lab coat while sharing the lessons of Advanced Chemistry with his students. Without forewarning, Mr. Ferri would oftentimes freeze in his tracks and present a positively devilish face to the class. I quickly learned what that meant.

Mr. Ferri would then rush to the cabinet at the front of the classroom and retrieve Cuddles to let us know that a pop verbal quiz was about to begin! Cuddles was a child's toy, and a particularly silly looking one at that. Cuddles was a small, colorful, cartoon character of a horse on wheels that stood maybe fifteen inches high. It was apparently designed for a child sitting on its saddle to push himself or herself merrily along the floor.

As nervous as I was as a student (I received the Nervous Nelly Award upon my graduation from elementary school), it was pretty hard to get terribly anxious about pop quizzes with Cuddles staring down at you from the teacher's lab bench. If you answered Mr. Ferri's question with the correct response, you were required to walk up to Cuddles and pet it on the head. A wrong answer meant you had to go up to Cuddles and pet it on its backside. It was pretty difficult for us to hold onto feelings of disappointment or frustration resulting from answering a question incorrectly while you were engaged in the rather undignified act of petting the tail end of a toy pony and giggling uncontrollably.

As memorable as Cuddles was and is, there was something even more fascinating that happened in class one particular day. Mr. Ferri presented the results of the famous double-slit experiment that was designed and used to demonstrate the wave-particle duality of photons. I'll cover the double-slit experiment in great detail later in the book, but allow me to explain the wonderfully mysterious result of two-slit experiments that just stuns everyone I know (especially me!) every single time they hear or think about it.

The double-slit experiment involves a simple experimental apparatus that consists of a panel placed between a light source and a screen upon which to project the light. Two precisely crafted parallel slits are cut into the panel to allow light to pass through. If the light source is illuminated, then the well-behaved light waves pass through the slits, and an interference pattern is produced on the screen. If we make any attempt in any way to figure out through which slit particular photons of light pass, the interference pattern disappears and is replaced by a much less interesting pattern. There's simply no trace of the interference pattern any longer.

This is fascinating, sure, but it gets even better. Let's return to the experiment and not attempt any longer to measure the path individual photons take through the slits. This time, however, let's just send out one photon at a time. As we do that, we observe the location on the screen where each photon lands. We watch as more and more photons illuminate the screen and eventually reconstruct the interference pattern we observed previously when we were dousing the screen with loads of light waves. The question was, "With what, exactly, is the photon interacting to make an interference pattern when only a single photon is in transit between the light source and the screen at any given time?"

Mr. Charles Ferri
Adv. Pl. Chemistry, Chemistry

Charles M. Ferri and Cuddles.
Source: Comet 1977 Yearbook

That lesson sent a chill down my spine and started a thirty-four-year-long thought experiment that culminated in the writing of this book. I spent much of the intervening time reading books on various subjects (quantum mechanics, cosmology, cosmogony, relativity, the physics of empty space, free will, classical philosophy, natural philosophy, science philosophy, the Anthropic Principle, the latest deep space experiments and observations, the nature of time, and other topics) in what had appeared to me during that time to be a futile attempt to get my arms around a rational explanation for the double-slit paradox. But this was all so that I could finally get a good night's sleep and find other topics to talk about at parties.

Try to picture this, if you can. After all, I was completely consumed with the double-slit paradox. I actually remember, quite vividly, being on a date with the stunning daughter of my dentist. Things were going pretty well, at first, while I focused on learning about my date and her thoughts and interests. Unfortunately, the list of her thoughts and interests eventually came to its end and she inquired about mine. The rest of the date bombed. Oh, well.

Over the decades, I learned about other paradoxes and mysteries that science appeared ill-equipped to overcome. I managed to preserve and stoke my interest in cosmology and quantum physics while embracing the notion that there's a time and a place for all things. Dates, apparently, were not the best time or place for discussing such topics, even if I personally found their mysteries to be exotic and romantic.

Eventually, I met my true life partner and I fully expect that we'll remain together for the rest of our days. I knew it at the start of our very first date. Early in our conversation, she asked me what university I had attended. "Johns Hopkins," I responded, full of pride and counting very much on impressing her. Without so much as a moment's hesitation, she unflinchingly burst my bubble in one fell swoop when she responded coolly, "Oh, that was my 'safety' school. I went to MIT." Sure, my pride was instantly devastated on one hand. On the other hand, I was also completely pumped at the prospect of a partner with whom I could discuss my thoughts about the nature of the universe! I was in love. We were married a year later.

With my wife's encouragement and support, I delved more aggressively into the paradoxes and complexities of our universe. And now I had a partner with whom I could go to bookstores and look for more content to consume. It's true that at bookstores my wife went to the sections that housed books on permaculture or cooking or knitting or travel and I went straight to the physics and, occasionally, the science fiction section, but it's also true that we'd meet up in the poetry section at some point to read love poems to one another. In any event, I continued with gusto on my quest to learn what I could about the directions that modern researchers were taking to try to tackle the mysteries that consumed me and so many others.

During the 1990s and the 2000s, it seemed like a roller-coaster ride as I kept up with new lines of cosmological research and reasoning that were at first pursued with gusto and then left to languish—some being abandoned altogether. Scientists and theorists alike were positively electrified by the prospects of multiple dimensions.

There was great potential for what might be learned through experiments at the Large Hadron Collider. Hopes had been running high for new insights into the nature of the universe when the Large Hadron Collider had begun operation in September 2008. However, major repairs were required shortly afterward. The repairs not only delayed experiments for more than a year, but the collider was operated more conservatively to prevent further repair delays. The science community had to temper its hopes for speedy new cosmological insights and muster its patience again. While I was writing this book, the collider had only been operated at 3.5TeV, an impressive, record-setting value, but still only half the energy for which it was designed.

In July 2010 my wife attended an international fiber and textile conference in Albuquerque. I was left largely on my own for nearly three days! Having that sort of freedom presents all sorts of opportunities to different people. For me, it was a perfect time to pick up another book about the nature of the universe and consume it.

This time I decided to leave the relative comfort of the traditional cosmology genre. I was feeling pretty adventurous, so I chose a book to read which was becoming quite popular: *Biocentrism* by Dr. Robert Lanza. It was nearly impossible to avoid the excerpts from his book that were cropping up all over the Internet, so I gave in and downloaded a digital copy of the book in the privacy of my home, where no one could observe me buying or reading such a "New Agey" sort of cosmology book.

Now, mind you, my motivation was not all that pure. It was my intention to read the book so I could more effectively refute it like a dedicated physicist was expected to. I considered myself to be firmly and exclusively entrenched in the cosmology camp embodied by the likes of Stephen Hawking, Lisa Randall, Brian Greene, and Edward Witten. After all, you know what Julius Caesar said: "Keep your friends close and your enemies closer." I needed to know what the other camps were thinking so I could better defend my position. It became necessary to penetrate the biocentrism camp.

The book had the completely opposite effect on me. The views that Dr. Lanza presented in his book changed my thinking in ways from which there could never be retreat. Before I had actually finished reading the book, it was abundantly obvious to me that Dr. Lanza's writings provided me with the pieces of perspective that I had been desperately seeking. Everything I had learned and everything I thought I knew just exploded in my mind and, as possibilities first erupted and then settled down, a completely new understanding emerged. The information I had accumulated in my mind hadn't changed, but the way I viewed it did—in a really big way. Finally, a single, coherent picture of our universe resulted with the potential to overcome the paradoxes, complexities, and inconsistencies of the theories of the universe that prevailed up until this time. This framework became part of my very identity.

Through the incorporation and subsequent refinements of the new thoughts that resulted, I achieved a new perspective with which to view all of the information I had at my disposal from a lifetime of collecting. (Actually, "hoarding" might be a more descriptive word to use. I only acquiesced during 2010 to my wife's ongoing demand that I give away my science fiction library if I wanted to collect more books, and I agreed!) Nonetheless, a new philosophy of our world unfolded in my mind. It appeared to have unified both the spectrum of fundamental questions posed by the early Greek philosophers and the paradoxes that modern science appeared unable to overcome.

Questions as seemingly disparate as these became immediately and deeply intertwined:

◊ "Who am I?"

◊ "Is reality an illusion?"

◊ "Into what is the universe expanding?"

◊ "What is time?"

◊ "How many dimensions are there?"

◊ "How can light travel through empty space?"

◊ "What is space?"

◊ "How in the world can single photons in a double-slit experiment show a classic wave interference pattern?"

In a universe governed by simple laws that bind the relationships among all things—living and inanimate, large and small—these questions must necessarily be entangled and inseparable. Insights arising from answering *any* of these questions, or countless other questions just like them about ourselves and our universe, *must* lead to insights involving *each* of the others. I choose not to represent this assertion as a personal opinion, but rather as a fundamental truth: the First Principle among all first principles. From this First Principle, I will share with the reader the alternate perspectives about the information collected by philosophers, scientists, and others throughout human history about our universe. I will also share perspectives about the theories that enabled me to frame a new model involving a perspective only mildly different from those usually accepted as scientific truth, albeit with sweeping consequences.

When combining these new, personal insights with the already existing and extensive body of accumulated scientific evidence, I discovered that the answers to each of the paradoxes were actually already there, in a sense, time and again. Scientists using the scientific method faithfully did their jobs well and collected all of the key information. However, the compartmentalization of knowledge by each discipline appeared to me to have combined with faulty, incomplete, and inherently biased human assumptions and interpretations in such a way that they conspired to lead our scientific thinking astray. The acceptance of potentially flawed scientific interpretations appeared to me to have led us to the very dead ends from which paradoxes arise.

Paradoxes don't emerge because the universe is inconsistent. Instead, paradoxes emerge when our assumptions, interpretations, and understandings are inconsistent with the true nature of the universe. The emergence of a paradox is the signal that informs us we've established a faulty assumption or made an erroneous interpretation. Through an iterative process of isolating and adjusting the faulty assumptions and interpretations, I discovered a perspective from which the paradoxes simply dissolve. It would be far more accurate to say that the paradoxes simply don't appear. Indeed, paradoxes can *only* exist in our minds.

It's true that the scientific method represents a best effort at a methodical, deliberate process that depends on the skepticism and critical review of peers from within the scientific culture. It precedes the acceptance and assimilation of new thoughts and concepts into the growing body of scientific knowledge.

I would never mean to imply that the process should be abandoned or that there's any better method that ought to replace it. But it's important to consider that science is, by its very nature, a flawed process with the potential to incorporate flawed assumptions as truth. It has the potential to build upon these flawed assumptions to construct flawed conclusions, especially when science is performed through a complex and increasingly ambiguous scheme of compartmentalization into subspecialties.

Such a system has the potential to ignore the meaningful contributions, thoughts, and comments of other researchers of whom they may be entirely unaware or which, even worse, they may deem irrelevant or unimportant. Consider the very real possibility that the best, and perhaps only, indication that non-truths have invaded the scientific method itself, as well as the body of scientific knowledge it produces, is the emergence of paradoxes.

I'm suggesting that the scientific method needs to be honed and adjusted in some fashion by the very same methodical process scientists use in adding new content and understanding to the body of scientific knowledge. The scientific method needs to be applied to itself in order to provide a continuously improving scientific method upon which we can increasingly depend. Neither our scientific method nor the body of scientific knowledge we accept as truth, nor our understanding of the world, can be accepted as static if truth is seriously our goal. The quest for truth cannot be separated from the need for continuous re-examination of the methods, assumptions, and interpretations upon which our perception of knowledge is based.

By the time I completed my reading of *Biocentrism*, the potential for science to build a house of cards or, more aptly, a loose affiliation of numerous such houses, seemed evident. While our technologies and our gadgets have continued to improve at an ever-quickening pace, our understanding of the universe and the continuing lack of respect we show toward our fellow life forms (for example, each other) has been striking in its stagnation.

True metrological value of any scientific measurement is elusive due to the presence of noise. This is a well-established tenet of science embraced by every student and practitioner. Science strives to reduce the margin of error inherent in all measurements through schemes such as repetitive measurements, the use of statistics, the elimination of sources of noise, and improvements in the technology of metrology. Indeed, one of the aims of science is to reduce the margin of possible error for the measured value of each of the universal constants in order to more accurately reflect the true nature of our universe.

Over time, the scientific method itself appears to be capable of accumulating margins of error in the form of incrementally faulty conclusions, which then themselves become the faulty foundation for the assumptions established by subsequent scientifically based premises and theories. The paradoxes we observe may very well be *created*, in whole or in part, through the integration of these faulty conclusions in the scientific system over time.

In keeping with the scientific method, I advance the following hypothesis: Paradoxes can occur through the accumulation of the margins of error relating to incrementally faulty conclusions that contaminate the scientific body of knowledge.

The paradoxes that haunt science and which have haunted me for thirty-four years simply evaporate when the faulty "interpretation" content is identified and then eliminated or adjusted based on newer findings. The information I accumulated over my lifetime and the thought process I implemented provided me with a platform by which I felt I could reasonably perform this process—keeping in mind that I am just another imperfect, flawed observer.

Suffice it to say, a compelling and singular perspective emerged from this exercise in which the existence of known paradoxes collapsed through a particular set of adjustments to scientific interpretations. At the same time, they remained thoroughly consistent with all scientific observations. There may be many such perspectives, but I had formulated at least one of them. Through the logic I used, this single formulation represented the proof needed to validate the existence of such formulations and thereby confirm my hypothesis. In a single flash, on an evening just a week before I embarked upon writing this book, I had the epiphany. It enabled me to finally overcome the ghosts of paradoxes past and present with which I've been wrestling for so long.

I thought that I had slept poorly when I was just pondering paradoxes. Believing that I had contributed to the prospect of overcoming them made my sleep go from bad to terrible.

I suppose that the result of reading *Biocentrism* was yet another example of something I've been fond of saying. When asked about the possible outcome of events in the future, I've always said that the only thing I know for certain about the future is that it will be far different from anything I predict.

What made this epiphany particularly unique and memorable, however, was the realization that there was, in fact, no future or past, except within our minds.

If you thought the stories about Cuddles were strange and entertaining, then just keep reading.

Scott M. Tyson
Albuquerque, New Mexico
May 1, 2011

Introduction

B eginning with the pre-Socratic philosophers, thinkers focused on the components and elements they perceived in the universe. To these philosophers, the universe consisted of people, nature, the Earth, the Sun, the Moon, the stars, and empty space.

Newton shared much of this worldview first established by the likes of Socrates, Plato, and Aristotle. By the twentieth century scientists and philosophers were convinced that the universe consisted exclusively of what they could measure—meaning ordinary, luminous matter. They became convinced through interpretation of measurements that the universe was expanding. By the early twenty-first century, our instruments provided us with data by which we surmised that regular luminous matter constituted only 4 percent of the mass-energy in the universe.

Enter dark matter, observed obliquely through the motions of rotating distant celestial bodies. Astronomers and astrophysicists determined that dark matter constituted 20 percent of the mass-energy of the universe—five times the ordinary matter observed and measured by all previous human attempts. Now, enter dark energy, again observed obliquely by the apparent expansion of the universe, a relentless negative gravity force driving matter, dark and light alike, away from one another, expanding the supposed edges of our universe at speeds far exceeding that of light itself.

Yet empty space represents the single largest component of our universe, far eclipsing any of the other features we observe. Even supposedly "solid" matter contains less than one part of matter by volume for every 500 trillion parts of empty space. Incredibly, little is known about empty space and the role it plays in the physical phenomena we observe in our universe.

It's All in Your Mind

Truth can be elusive. Just ask anyone involved in quantum mechanics, the judicial system, law enforcement, management, or parenting.

Truth can be particularly elusive in the modern educational system, especially from the perspective of students. In fact, education largely seems to consist of pumping the young mind full of age-appropriate facts, figures, and explanations portrayed as truth, only to be followed by the next teacher's assertions that anything the student had previously learned was wrong and needs to be replaced by a newer, better version. For those of you whose educational paths included the study of physics, there's no better example than the difference between Newtonian and "Einsteinian" physics.

As astounding as it seems, there have only been three schools of physics in all of human history prior to the emergence of quantum mechanics. The first school of physics was established by Aristotle, the essence of which was captured in his book, *The Physics*, which was really more along the lines of metaphysics and philosophy despite its title. The second school was established by Sir Isaac Newton during the seventeenth and eighteenth centuries and is often referred to as "classical mechanics." The third school was established as a result of Einstein's work in relativity performed at the start of the twentieth century; scientists and experimentalists confirmed his theories throughout the remainder of that century.

My early life as a student was spent learning and coming to believe that the equations derived by Newton and his predecessors, colleagues, and followers were accurate and real. Not one of my middle or high school educators ever happened to mention to me that classical mechanics was ultimately inaccurate and that it only represented a special case useful to our everyday existence.

Here I was, attending high school in the second half of the twentieth century, when Einstein's relativity was already embraced as a more accurate representation of our world, and no one bothered to tell me that I would soon encounter a new and improved version of physics! With all due respect to my teachers, of whom I remain mostly fond and appreciative, I don't believe that they were trying to pull the wool over the eyes of a naïve, eager, youthful student. I'm perfectly happy to give each of my teachers the benefit of the doubt that either (a) they themselves were unaware or unappreciative of the differences between the two schools of physics, or (b) that they were keenly aware that the next step that each student would encounter in the educational process would involve their next teacher instructing them to forget much of what they had learned previously and held as scientifically sacred. From my perspective, there was something akin to betrayal in this process that increased my skepticism over time of the "knowledge" that my instructors were passing on to me as accepted truths.

Further study of science in high school, college, and afterward required that I renounce that pesky simplification of the Newtonian world and replace it with the more physically accurate model of relativity. I was all too happy to do that because relativity seemed totally cool while classical mechanics seemed dull and lifeless. The change made me feel "space-aged."

It didn't take too long the second time around to realize—this time somewhat in advance—that relativity itself also represented a simplified, not-entirely-correct approximation of the universe, whether my teachers were willing to admit it or not.

My experience was that teachers just absolutely adored presenting their students with the paradoxes inherent in the models they teach; relativity was certainly no exception. I believe there are two good motivations teachers have for doing this and they both have to do with being dedicated educators. The first is that it must be the ultimate thrill as a teacher to present the really juicy paradoxes before another instructor has had the chance and then watch their students' jaws drop and slam against their desks. The other, I think, is the deep-seated, nearly spiritual hope that he or she will motivate their students to confront and overcome the challenges inherent in paradoxes, thereby assuring their own place within the proverbial circle of life and the unfolding of human destiny.

Then there was quantum mechanics, and all bets about everything I thought I understood were off.

Everything You Have Ever Learned Was Wrong

That's right: Everything you have ever learned was wrong. It's true in a very real sense. I need to convince the reader of this before taking too many steps backward and then, hopefully, forward in a boldly adjusted and new direction regarding our understanding of the universe.

I'm terribly sorry to be the bearer of this news, but here it is: Humans appear dramatically limited in what they can hold at one time in their thoughts, so they make models that embody the approximations they have chosen to embrace. (A word to the young: These limitations don't improve with age, so make the best of it while you can!) We don't necessarily realize consciously when we do this, but it appears to be a natural consequence of being human. The impact of the assertion that everything we've learned about the world is inaccurate should not be too terribly surprising or devastating, even if we participated in such an imperfect learning process without being keenly aware of it. Most of us, it would seem, have picked up on at least some clues about this state of affairs throughout our lives.

So, the first step we'll take in this process is to explore and attempt to recreate certain key elements of classical logic and the somewhat tacit, underappreciated role it plays in the development of the personal framework in which we view, understand, and interact with our world.

Here's a short example: Some of us were taught and accepted as truth the theorem from Euclidean geometry that parallel lines meet only at infinity. Here's the kicker: They really don't. This assertion is just a concept that exists solely in our minds and nowhere else. If a flat, infinite, Euclidean plane actually existed, parallel lines would remain parallel—both near and far—and never, ever converge. Period.

Through this theorem, we're being asked to combine wildly disparate abstract concepts (two straight lines placed a finite distance apart that extend to infinity in each direction sitting on an infinite plane) in the only place where such abstractions exist and can be stored: our minds. Furthermore, the only place where we ever think that we see parallel lines converging is within our perception of the visual perspective of distance, just like when we stand in the middle of a long road and look out in the distance; that also *only* exists within our mind. There is no perfectly flat Euclidean plane; there are no perfectly straight lines; and it is highly unlikely that any lines of any sort actually extend infinitely through either the space or time of classical mechanics, or the space-time of relativity.

At the risk of sounding like I'm actually heaping some sort of praise on the movie, *The Matrix*, with Keanu Reeves, I offer the following scenario for digestion. This idea reminds me of the scene in the Oracle's apartment in which a stoic child who is apparently bending a spoon with his mind informs Neo (played by Reeves) that the key to bending the spoon is acceptance of the realization that the spoon doesn't actually exist.

Let's consider another example of something we might often take somewhat for granted: music. Most of us know about Pythagoras, the Greek mathematician who is credited with development of the Pythagorean Theorem: The sum of the squares of the lengths of the sides of a right triangle on a flat plane equal the square of the length of the longest side called the "hypotenuse" ($a^2 + b^2 = c^2$). But Pythagoras did far more than develop a theorem with which to task students with endless homework problems; he succeeded in providing us with the first musical scales.

At times it seems difficult for me to grasp the theory underlying music since my own musical skills are limited to opening the Rhapsody app on my computer or phone and then selecting the artists I want to hear and the volume at which I want to listen. Pythagoras painstakingly determined the relationships between sounds of different wavelengths and constructed a variety of musical scales that remain in use today and will likely remain for the foreseeable future. In essence, Pythagoras developed a means by which people could distinguish sounds (acoustic wavelengths and their combinations) that are music from those that are not. That seems amazing to me!

I'm grateful for Pythagoras' achievements and contributions, especially since

I have a deep fondness for music. It provides enjoyment, a therapeutic escape, and a means by which to feel emotional connections to other people and circumstances. Yet music doesn't exist in the universe, either. It *only* exists in our minds. (Dr. Lanza does an excellent job of explaining this phenomenon in greater detail in *Biocentrism*.) And, if that weren't already enough to consider, then consider the fact that Pythagoras created musical scales that are associated with human emotions! Particular relationships between air vibrations of one wavelength and another are intricately entangled with human emotions. How could that possibly be? Regardless, both music and the experiences it creates within us exist nowhere else in the universe but in our minds.

The Quest for Truth

Before we take a single step on our ambitious journey to construct a new model of the universe free of paradoxes, we really have to look behind us and understand the journeys undertaken by our ancestors and predecessors—in other words, understand *their* world. This is where we find both the foundations of logic and the roots of some of the early questions that remain unanswered (or ignored) today. I maintain that while the foundations of logic became the basis for today's scientific method, I equally maintain that the unanswered, insufficiently answered, and ignored questions became the very basis for the paradoxes we observe today.

A truth seeker is limited by the body of knowledge available to him or her at any particular time. Truth seekers have little choice but to accept the prevailing body of knowledge as pure and uncontaminated. Truth seekers are particularly and keenly aware that perfection is unachievable; therefore, the prevailing body of knowledge contains inaccurate and incomplete content, since they themselves seek to make contributions to advance the state of knowledge and understanding. Even at its very best, the existing body of knowledge may include, at least in part, *inaccurate* or *incomplete* assumptions and conclusions accepted falsely as truth.

It seems, then, perfectly reasonable to consider that any truth seeker's assumptions and conclusions might be different with advancements in tools and/or knowledge. What might Aristotle, Plato, and Socrates, and their predecessors and students of their time period have deduced if they had telescopes, atomic-force microscopes, and magnetic resonance imaging tools, let alone cell phones, computers, and the Internet through which to communicate? By the same reasoning, so might Newton's and Einstein's propositions, interpretations, and conclusions have been different, especially had their assumptions about the universe been in any way different from what they were.

The scientific method, in fact, requires that new scientific theories and explanations be constructed upon the existing, and possibly (or even likely) flawed foundation represented by the then-current scientific body of knowledge. The scientific method is designed to build methodically and deliberately upon the logic, propositions, conclusions, and interpretations that have been reviewed and embraced as truth by scientist-peers at each point in time throughout the process and accepted as rote by those that follow. In essence, the current scientific method requires that new theories and understanding be based on a foundation that can *only* be flawed.

Like parallel lines, music, logic, and paradoxes, truth itself exists nowhere else in the universe other than the mind. Consider logic as *the process that we have chosen to accept that attempts to lead us toward an understanding of truth*, and truth as *the state within our minds when we believe that our perceptions have been synthesized into a sufficiently accurate, internally consistent portrayal of our world, or some portion of it, at some moment in time.*

The available literature covering the topic of logic is staggering, and the spectrum of views cannot be replicated faithfully here in any sort of succinct form. Other discussions on incremental differences between individual perspectives on logic are really quite fascinating and seemingly endless and have proven helpful to me as an antidote in overcoming the sleeplessness resulting from my thought experiments about the universe. (This comment should ensure the avid scrutiny of my propositions by the community of logicians and philosophers. Remember, scrutiny is a good thing and should probably be applied in ample doses to new thoughts and old interpretations alike.)

In short form, reasoning is the process that leads us toward insights involving truth. Logic is the method by which we attempt to reason. Reasoning involves the establishment of assumptions, the collection of new observations and/or thoughts, and the synthesis through logic of new understandings that lead to truth. Each element of this process—namely reasoning, assumptions, observations, thoughts, logic/synthesis, understanding, and truth—exist nowhere else in the universe but in our minds. From the perspective of the observer, then, the items on this list are not properties of the universe; on the contrary, each is a property belonging only to the self.

Consider that paradoxes "exist" as a result of incomplete and/or inaccurate knowledge. It would therefore be paradoxical itself to consider the prospect of overcoming scientific paradoxes through the use of existing scientific knowledge and interpretation. Chew on that!

It seems to me a very sticky dichotomy exists that is built into our beloved scientific method. It goes like this: No self-respecting scientist would consider the prospect that science is *perfect and infallible*, yet no self-respecting scientist would question the foundations of their assumptions if they retrieve those assumptions from the existing and accepted pool of understanding called "scientific knowledge." The flip side of this coin is that no self-respecting scientist would permit as valid the use of assumptions that don't derive from the current pool of "scientific knowledge or understanding." Sorry, but we just can't have it both ways. Put simply, it's illogical. Put another way, it's paradoxical.

I'm going to state it again, just in case the reader nodded off: An imperfect (a.k.a., flawed) scientific process ought to lead to the observation of paradoxes. I can think of no alternative.

Continuous retrospection, re-examination, and re-interpretation may be the only process by which scientists can identify and scour out the contaminating content. If we're willing to accept that premise, then we must become familiar with the path that led us to the current state of scientific knowledge.

Enough of what I think for now. Throughout the book, we will review some of the steps taken by early philosophers that ultimately relate to the theories I propose later in the book.

Section I

Sensation, Observation, and Interpretation: Acquiring Perspective

Chapter 1

A Few Things to Think About

Since the dawn of humanity, perhaps even since the very rise of consciousness, humankind and its earliest ancestors tried to make sense of the world around them. They formed mythologies and religious beliefs in an attempt to explain the phenomena they observed. That world must have been a terribly frightening place. Our ancestors huddled in caves and trees during storms and eclipses as they struggled to survive the events that nature threw their way.

As humankind progressed and civilization took hold, people began to have the opportunity to sit down and consider their circumstances. They began to ponder their world in more rational and deliberate ways as civilization and society began to provide for the basic necessities of daily living. This movement forward spawned the rise of logic and philosophy to begin the process of replacing the explanations provided by their mythologies and religious beliefs with factual, physically based, internally consistent comprehensions. Logic and philosophy ultimately gave rise to science and the scientific method as means to better understand our world.

The questions that we ask today have not changed in any significant way over time, even as our deepening thirst for answers to fundamental questions has motivated us to develop ever-improving and profoundly more capable technology. The technology is only a means by which we probe and measure our environment. It neither alters the basic questions we ask, nor has it alone substantially altered our view and understanding of the world. Sure, technology is needed to test our scientific hypotheses and produce evidence of their trueness or falseness. Yet, as humans, it is difficult to completely separate our observations and the scientific data we produce from our internal biases that color our interpretations in ways that vary from the subtle to the not so subtle. As a result, our collective scientific interpretations accumulate our biases and begin to expand the possibilities that our interpretations become disjointed from the data that produced them.

Some early Greek philosophers believed that abstract logic and reason were the only means by which to understand our universe, to avoid the sensory illusions that led to faulty interpretation.

Over the last few decades, there has been little fundamental progress in understanding the universe. Not coincidentally, over this very same time, the diversity of thought regarding the nature of our cosmos has not expanded significantly and the cost of technology-based experiments to probe our hypotheses has grown exponentially. I don't question the need to accumulate empirical evidence for carefully controlled technology-based scientific experiments, since our approach to science demands this. However, cosmological research seems to be serving technology as the master instead of the other way around. It seems as though technology has taken on too much significance and has possibly become a costly distraction, both in terms of precious (and currently dwindling) funding and in terms of thought.

Our resources have become nearly singularly focused on demonstrating a narrow cosmological perspective that has solidified to the point of achieving religious doctrine. Yet, the current understandings and models of the universe must be accepted as containing some combination of incomplete and flawed content as evidenced by the large and rapidly growing body of scientific paradoxes they produce.

As an example, nearly all cosmologists and particle physicists believe in the ultimate unification of the four known forces (gravity, strong nuclear force, weak nuclear force, and electromagnetism). Our science, math, and technology have allowed us to unify three of the four forces; gravity remains the outsider. Scientists agree that no empirical evidence exists upon which to believe that the four forces will unify into a single Theory of Everything (TOE), yet nearly every scientist will strenuously and ardently defend his or her unfounded assertion that all of the forces will unify indeed.

Let us remember that strong beliefs for which no empirical evidence exists is the very way in which we, scientists included, define religion. Scientists are not above religion and, although some would provide ample arguments why their belief in unification does not constitute religion, at the end of the day, it is religion, pure and simple. I don't mean to insult the scientific community in any way by these comments. Rather, my hope is to point out that, although we develop and follow different paths and methods to achieve our goals, in the end, we are human; it's implausible for us to interact with our world in any way other than through the limitations of the human condition. We are human, after all, no matter what we do.

Please don't get the impression that I harbor any sort of criticism toward the scientific method and the way it is practiced today. And I certainly hope that I don't provide the impression that I myself subscribe to a different outcome involving the four fundamental forces. Let me be clear: I deeply subscribe to the notion that the four forces do unify into a single entity. In fact, I'd go further in stating my belief and my understanding that everything we observe, experience, and measure is ultimately comprised of a single phenomenon, and that there is but a single underlying reality.

We've come to understand that our current model of the cosmos—our understanding of the universe—produces a staggering number of paradoxes. Consider the fact that even a single paradox provides ample evidence that our model is either incomplete or flawed, or both. Consider also the fact that, as I have stated earlier, paradoxes don't exist in the universe. The only place in the universe where paradoxes about the universe exist is inside our minds.

With this thought in *my* mind, I decided to embark on a project to review as much of historical science and philosophy as I could. I attempted to understand how particular events might have contributed to leading our interpretations astray. I asked myself the following questions:

◊ Is it possible to identify a small set of scientific events for which the prevailing interpretations could be subtly altered and yet remain fully consistent with scientific data that could result in the elimination of all known scientific paradoxes?

◊ With the power and benefit of hindsight, is it possible to identify and eliminate faulty human biases and their resulting flawed interpretations to produce a model of the universe in which paradoxes simply never arise?

After the decades of personal study and research I dedicated to this project, I now believe that it is possible to develop a framework by which to understand our universe and to see it free of paradox. I will propose in this book a single framework that accomplishes this, although I believe that other frameworks exist that might provide similar, if not identical, results.

In order to effectively present this framework to you, it's necessary to review the history of philosophy and science that brought us to our current circumstances. For this project, that review begins with the pre-Socratic philosophers: the Atomists. These Atomists, along with their students and followers, asked the questions that actually led to the rise of the modern scientific method. Many of the questions they asked remain unanswered even today, and the paradoxes they described continue to nag at our collective consciousness. These questions and paradoxes provide us with a common and meaningful understanding of humankind's starting point toward questioning and exploring our world.

I hope to share with you throughout this book a perspective that remains true to all of the rich scientific data that has been accumulated. I will attempt to share with you a framework in which I will propose small changes—really only subtle tweaks—to only a handful of scientific interpretations that were presented and accepted as rote by the scientific community over the last 350 years or so. When these adjustments are folded into our understanding of the universe, we see that many, if not all, of the paradoxes that arise from the commonly accepted view of the universe simply evaporate. Rather, the paradoxes just don't appear.

I submit this framework for the reader's consideration as perhaps just one example of a potentially large number of such frameworks that stand to eliminate the paradoxes.

Do photons "carry" light? If so, how? Or must we say photons "ARE" light?

Chapter 2

The Double-Slit Experiment

Perhaps the most well-known example of a modern scientific paradox is the double-slit experiment. The double-slit experiment is a simple laboratory apparatus consisting of a coherent light or wave source, an opaque panel through which two precisely formed slits have been cut, and a screen on which to project the light streaming through the slits. The experiment was first proposed and performed by Thomas Young in the early 1800s to examine the wave-particle nature of light, even though he performed many of his experiments with water.

Let's examine this wave-particle nature of light a bit more closely. Attempts by researchers to study light demonstrated that light exhibits both wave behavior and particle behavior—but not at the same time. For example, if the researcher tries to demonstrate that light is a wave, then the experimental results clearly demonstrate that light is a wave. Yet, if a researcher attempts to demonstrate the particle nature of light (namely that light is carried by very small particles that we call photons), then the results clearly demonstrate that light consists of particles. *?*

How in the world can that possibly be, you might ask? That is a reasonable question, but let's take a little step backward for the moment so we can better understand the mood and expectations of earlier times.

In our macroscopic world, we've become very accustomed to everyday items possessing one set of attributes or another. We're accustomed to objects being cold or warm, hard or soft, smooth or rough, blue or red—or some other color. The vast majority of our experiences with the world around us allows us to characterize items in very definite ways with which we have become really familiar.

We don't expect any object, at any one moment and under any single set of circumstances, to be describable by two exceedingly different descriptions. For example, we wouldn't expect to describe an ice cube as both cold *and* warm, or as both rigid *and* malleable. Yet, if the circumstances change (think of this as changing our experimental setup), it becomes possible to apply wildly different descriptions to the same single object.

Certainly our description of ice differs greatly from our description of water and our description of water would surely be different from our description of steam. The most basic nature of water hasn't changed as we've changed the conditions for our observations. That is to say that, under any of these circumstances, water remains comprised of molecules with each molecule consisting of one oxygen atom bonded to two hydrogen atoms, yielding the familiar chemical expression "H_2O."

So, as you begin to ponder this example, I'm sure that countless examples enter your mind of everyday objects whose descriptions may vary greatly if you alter one or more simple conditions. Even if we take a single item like water, without shifting it between different phases of solid ice, liquid water, and gaseous steam, we find that a single description might become difficult. For instance, it seems like water is particularly easy to move if we stir sweetener into our iced tea or coffee, but that it's terribly resistant to movement if we try to sprint through a filled swimming pool.

As we spend time thinking about it, then, we see more and more examples of everyday phenomena that can each be described in various ways depending upon specific circumstances. The perceived properties of objects can also depend very strongly on the perspective of the observer. A window may seem very transparent to someone, while another person may be viewing the window from a perspective in which the sun is reflecting brightly off it. This second observer might describe the window as being highly reflective, like a mirror.

It begins to become clear that the description of an object depends upon many factors and that, as we alter the conditions to perform an observation, we may discover that a single phenomenon can be described in improbably and wildly different ways to different observers.

Let's go back to the double-slit experiment. At the time that the experiment was first performed, scientists did not even know for certain that electricity and magnetism were two faces of the same force that we know today as electromagnetism. Furthermore, the relationship between light and electromagnetism was largely unknown. It was James Maxwell, in 1864, who first established that what we refer to as light, or more specifically as "visible light," was nothing more than a single, very narrow portion of the electromagnetic spectrum, which also includes other phenomena that we refer to as x-rays and radio waves.

Okay, so Thomas Young performed his double-slit experiment with water waves by placing the panel with two precisely formed slits between his wave source and his back screen. The results he observed are shown in Figure 1.

Let's analyze these results. If we look at the pattern formed on the screen, we notice a few important features. Specifically, we see regions where the waves are very strong and other regions where the waves are very weak. This sort of pattern should seem somewhat familiar based on our everyday experiences. We have seen patterns at least reminiscent of this in nature, the most common example of which is the pattern made when dropping stones into a pond. The pattern has the appearance of waves interacting with each other. In some places the waves and their energy add to each other, resulting in regions that scientists describe as "constructive interference." In other places, the waves cancel each other out, resulting in regions of "destructive interference."

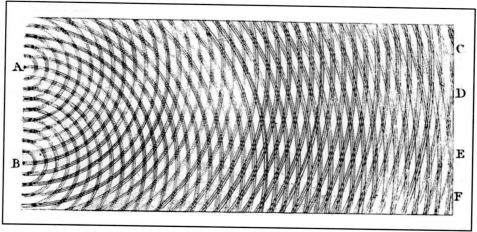

Figure 1. Thomas Young's sketch of two-slit interference, based on his observations of water waves. Source: Wikipedia.

It's common to perform the double-slit experiment using light, in which case we observe very similar results to those achieved by Young using water waves. (See Figure 2.) The bottom line after analyzing the results of this version of the double-slit experiment seems rather simple and straightforward. We see bright regions where the waves combine constructively and dark regions where the waves cancel each other out. Light is a wave. Wouldn't you agree?

I don't want to introduce too much of a distraction right now, but after observing the wave-like patterns of light on the screen, scientists also immediately think of the parallel with the waves produced by dropping stones in a pond. And they observed that the waves in the pond are carried by water. "What in the world can be carrying the waves of light?" they wondered. And so was born the theory of the Luminiferous Aether, which we will also discuss in greater detail later. By the way, no one has yet determined what is carrying the waves of light but you can bet we'll be talking a great deal more about that throughout the book. The implication from the results of this double-slit experiment is that light consists of waves.

Let's take another step back at this time and talk a little more about the double-slit experiment. In our last example we used light as the source. We know today that light, or what we more commonly refer to as "visible light" consists of electromagnetic energy. Electromagnetic energy covers a broad spectrum, including x-rays, gamma rays, microwaves, and radio waves. You might notice that many of these modern terms include the term "waves" in their names, as in "microwaves" and "radio waves." So today, we've come to think of light, and the broader concept of electromagnetic energy, as consisting of waves or at least that electromagnetic energy manifests itself as waves, some or all of the time.

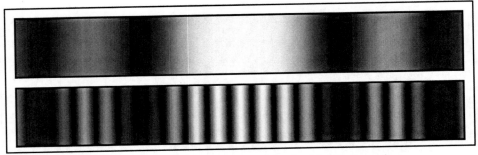

Figure 2. Single-slit diffraction pattern (top); double-slit interference pattern (bottom). Source: Wikipedia.

Perhaps you are also familiar with the term "photon." What, then, is a photon? The short answer is that the photon is the carrier of the electromagnetic force. (As you may recall, electromagnetism is one of the four fundamental forces—the other three forces being gravity, the strong nuclear force, and the weak nuclear force. We'll discuss each of these other forces in more detail throughout the book.) The photon consists of a very small quantity of electromagnetic energy, which scientists commonly refer to as a "quantum." These small bundles of energy interact with matter.

I'm not going to go into more detail about this just yet; I hope it's sufficient for the time being that we are satisfied with the concept that electromagnetic energy is absorbed and emitted by matter through these quanta of energy called photons. One more thing about the photon: it is common for scientists to consider photons as a special type of "particle." We'll see later on that scientists associate a particle with each force and these particles are called "carriers" of the force with which they are associated. These carrier particles are also called "gauge bosons."

Okay, let's get back to the double-slit experiment. As you will recall, for our last experiment, we chose to produce visible light and send it through the two precisely formed slits. Finally, the light landed on the screen where we observed an interference pattern associated with waves. The interference pattern formed bright regions where the light waves combined constructively based on the sum of the energy of some waves. The other darker regions are where the light waves combined destructively, where the energy of some waves canceled out some or all of the energy of other waves.

So far, I admit, this experiment doesn't seem terribly exciting or illuminating—pardon the pun—since most of us are familiar in one way or another with this phenomenon and the results are pretty much in line with our expectations. Here, and throughout the rest of the book, I'll present you with different variations of the double-slit experiment that provide astounding results that have eluded explanation. Some of these variations of the double-slit experiment have baffled scientists for nearly a century, while other variations of the experiment only became possible recently through advancements in technology and in the field of quantum mechanics. As you'll see, these double-slit experiments have blurred the boundaries between waves and particles, and have forced us to reconsider our notions of time and space as well.

You may be familiar with Dr. Richard Feynman, a famous and popular American scientist, teacher, and author of the twentieth century who played a crucial role in analyzing and interpreting the results of double-slit experiments, among his many other credentials and accomplishments.

Dr. Feynman became quite famous for the work he performed on the committee assembled to investigate the cause of the space shuttle *Challenger* disaster. When the time came for the committee to offer its findings, Feynman provided a very compelling demonstration. He soaked a sample of the material used to seal the solid fuel booster in ice water. He removed the chilled O-ring material and struck it with a hammer, shattering it. He provided a simple, irrefutable graphic demonstration that the O-ring material failed under the combined influence of cold, early morning launch weather and the extreme pressure and vibration produced by the solid fuel motors. Hot flames shot out the side of those compromised motors and toward the liquid fuel tank, which then became compromised; it exploded. His work truly underscored the nature of a national tragedy that could have easily been averted through more prudent engineering.

Dr. Feynman believed that the entirety of quantum mechanics, and perhaps, even a thorough and complete understanding of the nature of our universe, could ultimately be derived from the understanding of double-slit experiments. I hope to demonstrate that Dr. Feynman was indeed correct.

Allow me to describe another version of the double-slit experiment. This is the version where the weirdness begins. The results of this version of the experiment elude explanation by our contemporary scientific understanding.

Okay, so here we go. The experimental apparatus is the same. Again we will use a coherent light source, a film with two precisely formed slits, and a screen on which to project the light coming through the two slits.

Like the last time, we'll turn the light on and, once again, we see the expected interference pattern. But this time, we're going to turn the light source down to the point where we only have a single photon of light at any moment traveling between the lamp and the projector screen. Let me make certain to describe this really well. We have adjusted our experimental apparatus so the lamp produces only a single photon at a time. Furthermore, we have adjusted our light source in such a way that we ensure we don't emit another photon until the previous one strikes the screen and gets recorded. Only once a photon has struck the screen and its behavior recorded do we let the next photon rip.

Since this setup only allows us to observe the strike of a single photon on the screen at any moment, we'll need some way to capture the results over time so we can see if an interesting pattern emerges. (You'd be perfectly correct to guess that an interesting pattern emerges!) So, let's use an old-fashioned film camera focused on the projector screen and leave the aperture open for a while so each image of a photon strike adds to the previous combined image of photon strikes. We allow the results to be integrated for a long time, since we're using so few photons in our experiment. At some point, we decide to shut down the experiment and develop the film.

For you youngsters out there, there was a type of camera that wasn't digital and wasn't built into a mobile phone. In fact, we didn't even yet have mobile phones. These older cameras used something called "film," which you carefully placed in your camera while huddling in a dark place. You could only store two to three dozen pictures on your film roll before you had to change it. And, no matter how badly your photographs turned out, you had to pay to first develop and fix the negative images made in the silver iodide emulsion on the film and secondly to print the images from the negatives onto paper-based film. It used to be terribly frustrating to have to wait hours or days to receive your photos so you could enjoy the paltry number of pictures that turned out well.

In any event, I've probably built up too much anticipation during that side comment about film. I could have chosen to use a digital camera for the experiment but it would have been tougher to build up any anticipation.

Okay, your pictures were developed, printed, and returned to you, so you could begin your analysis of the experiment's results. Awestruck, you see the same interference pattern you observed in the original experiment. (See Figure 3.) How in the world can this possibly be? As far as we're concerned, any single photon never had the opportunity to interact with any other photons. So ... how in the world could we observe the same interference pattern as before? With what could we be "interfering?"

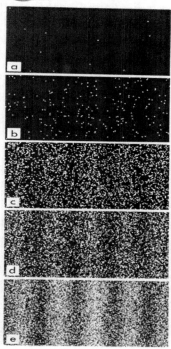

Figure 3. Results of a double-slit-experiment performed by Dr. Tonomura showing the buildup of an interference pattern using single electrons instead of photons. Numbers of electrons are 10 (a), 200 (b), 6000 (c), 40000 (d), 140000 (e). (Provided with the kind permission of Dr. Tonomura.)
Source: Wikipedia.

Now we really get to appreciate the first set of scientific paradoxes. This set of paradoxes consists of (a) the wave-particle duality of photons and (b) the interference pattern that results from the double-slit experiment. Allow me to assure you that there are many other variants of the double-slit experiment, and they get much stranger than even the results of the version I just shared with you.

There are a few other pieces of information I should probably share with you at this point. First, we aren't required to use photons in our experiment, but photons are particularly easy to produce. We could have chosen to use something other than photons, the carrier of the electromagnetic force. We could have used subatomic particles, like electrons, protons, or even neutrons. Neutrons are far more difficult to observe, since they don't possess a charge and they don't interact with matter in a way that's easy to detect and observe in real time. By the same token, we could have used atoms or ions (atoms with a non-neutral charge state), like hydrogen (a hydrogen atom with its one electron removed is just a proton, by the way), helium, carbon, or any other atom in the periodic table of elements. For that matter, we could have used molecules in our experiment, like table sugar, or water, or even a molecule of buckminsterfullerene.

Incidentally, buckminsterfullerene actually exists and is comprised of carbon atoms arranged in a geodesic-dome-like configuration reminiscent of Buckminster Fuller's architectural creations. In fact, that double-slit experiment has actually been performed successfully using these "Bucky balls."

Anton Zeilinger at the University of Vienna published the results of his 1999 experiment in which he produced the double-slit interference pattern using these relatively heavy Bucky balls.[1] Without belaboring this point much longer, the double-slit experiment can actually be performed using any object at all—even large, macroscopic objects like baseballs, microwave ovens, and automobiles—although the wave-like characteristic becomes far more difficult to observe and measure.

It turns out that a scientist named Louis de Broglie determined that all objects and phenomena exhibit wave-particle duality, meaning that at certain times an object or phenomenon exhibits wave behavior and at other times it exhibits particle behavior.

During the late 1800s, de Broglie is credited with providing an equation that relates an object's or phenomenon's wave behavior (its wavelength or frequency) to its mass or energy. "The de Broglie relations show that the wavelength is inversely proportional to the momentum of a particle and that the frequency is directly proportional to the particle's kinetic energy. The wavelength of matter is also called its de Broglie wavelength."[2]

Let's return to our double-slit experiment. Let's review what we've discussed so far. In the experimental setup, we produce photons from a light source and project them through slits onto a projector screen. Whether we send bunches of photons at a time or only a single photon at a time, we observe wave-like interference patterns when we record the positions of the photon strikes on the screen over time.

At this point, I'd like to make some statements:

1. Our science cannot currently and adequately explain the appearance of the interference pattern of the single photon variation of the experiment;

2. The interference pattern of the variation of the experiment in which we are sending out bunches of photons is indistinguishable from that of the single photon experiment; and

3. Something is causing the interference pattern in each case, whether we can explain it well or not.

Now, it's time to ask a question: As each photon passes from the light source to the screen, which slit does it go through? We think of a photon as a single entity, so this would lead us (rational people) to consider that there are really two possibilities—any given photon striking the screen must have passed through either the left slit or the right slit.

Let's modify our experimental apparatus so we can determine through which slit the photon passes. There are actually several ways to do this; researchers constantly develop and demonstrate new variants. One of the popular methods is to send out photons of polarized light. Polarized light consists of electromagnetic waves undulating in a single plane.

For example, let's say the light we produce is polarized in the vertical plane. Then we can place a filter in front of each slit and rotate the light a little clockwise or counterclockwise. In this way, we can tag each photon by shifting the polarization of the light as it passes through a slit in one way (clockwise) or the other way (counterclockwise). Finally, we modify our apparatus so we can observe the strike of the photon on the screen and measure its polarization.

We're all set. Let's run our new variant of the experiment. In this new variant, we will shoot our photons from our polarized light source, tag them by rotating their polarization one way or the other, and then we'll detect the photon strikes, measure the polarization of each photon, and determine through which slit each photon traveled.

Here we go. We've performed our experiments and there's no sign whatsoever of the interference pattern. We checked our setup to make certain that everything is correct, and we've run the experiment again and again, and we just can't observe that pesky interference pattern, so we figure the experiment is a bust.

Just as we're about to throw in the towel, we get an idea. Let's remove the polarization filters and run the experiment one more time. This time, however, we feel conflicted. Our old friend, the interference pattern, has returned so we take delight and comfort that we still are sane. At the same time, however, we find ourselves feeling terribly troubled by the fact that something as straightforward and simple as the presence of the polarization filters renders our experiment useless.

Let's take a moment to ponder this outcome a bit further. In retrospect, our experiment was not a failure. Remember that the scientific method doesn't guarantee a particular outcome no matter how expected it might be. Certainly, our result was not what we had expected, but it would be wrong to judge it a failure. Let's consider a possibility: Something about the attempt to determine the path of the photon eliminates the interference pattern.

As difficult as it might seem to accept, this is precisely the case. This version of the double-slit experiment always results in the loss of the interference pattern, regardless of how cleverly a researcher attempts to be in slyly determining the path of the photons through the slits. In fact, either attempting to measure the individual photon's path through the slits, or even simply the intention to do so by placing the filters in the pathway, eliminates the interference pattern, and replaces it with a pattern that can be produced by overlapping the results from two individual, single-slit exposures.

Clearly, there's more going on here than meets the eye!

This is perhaps a good time to introduce some other important concepts and learn some important perspective that will help us to set aside our frustration for a while and move us forward more productively.

The world of quantum mechanics is a strange one indeed. It may defy many of our everyday intuitions and experiences about the world in which we live. The quantum world is consistent in its behavior, and it is knowable—even if we only have an incomplete and possibly flawed understanding of it at this moment in time.

I'm going to describe a couple of fundamental aspects of quantum mechanics, and they are going to seem a bit peculiar at first. A fundamental element of quantum mechanics is that only phenomena or objects that can be measured are considered to exist. Another way of wording this is that only objects or phenomena that exist can be measured. As you'll see, this is slightly different than the previous statement. This is a concept to which we will be returning throughout the book.

Another fundamental element of quantum mechanics is that an observer is required to perform some sort of measurement of the object or phenomenon in order for the state of the object to take the form that can then be measured.

Each of these concepts can seem quite perplexing. Let's delve a bit more deeply into each. Admittedly, the first concept seems a bit easier to grasp. Throughout our daily macroscopic lives, we only consider as real those things that we observe through our perceptions of the world around us. If we observe something, we embrace the concept that it exists. We generally don't consider things we can't directly observe through our perceptions as things that actually exist. That's not to say that we don't accept as real those things that others observe and describe to us, such as viruses, bacteria, atoms, someone's child or spouse, and so on.

Our everyday experience trains us to think that if an object or phenomenon exists, then it exists whether our eyes are open or shut, and whether or not we experience it through our perceptions moment after moment. This expectation that we apply to the macroscopic world of our everyday existence just doesn't apply to the quantum world.

The science of the quantum world emerged during the first three decades of the twentieth century, and the following decades produced great strides. Brilliant scientists contributed to the development and subsequent debate involving quantum mechanics, the science of the quantum world. These scientists included Bohr, Rutherford, Heisenberg, and Einstein.

Quantum theory has enjoyed huge success. In fact, it is believed that every prediction ever made by quantum theory has been successfully demonstrated. It would be difficult to argue successfully in favor of making changes to the accepted tenets of quantum mechanics but, as we'll see later in the book, quantum mechanics and relativity produce different results when one is applied to the domain of the other. As a consequence, it becomes quite clear that one or the other is incorrect and/or incomplete or that both theories are wrong and/or incomplete. I'll provide a thorough treatment of these possibilities as we progress through the book.

In any event, I was mentioning that quantum mechanics only considers objects or phenomena that can be measured as existing. This is so fundamental to quantum mechanics that its assertion was established as a key fundamental tenet of quantum mechanics resulting from the meeting of the minds of that day and captured within something called the Copenhagen Interpretation.

The reason why this is so important is that phenomenon in the quantum world can actually exist in many states simultaneously. It is the act of measurement that causes the many states, or the many possible states, to collapse down to a single state: the one that the observer actually measures. Until that measurement occurs, the phenomena is said to exist in an indeterminate state, and the act of measurement forces this indeterminate form of the phenomena to collapse down to one single possibility. So, if there's nothing to measure, the Copenhagen Interpretation states that there's nothing that exists.

Conversely, when we complete our measurements or stop taking measurements, the phenomena returns to the indeterminate form. Taking this one step further, phenomena and other objects exist only when they are under observation and measurements are performed. As peculiar as it sounds, quantum mechanics says that phenomena and objects cease to exist in the convention sense when they are not being observed. It's strange to be sure, but it's true.

It's true in one sense. It's true that the object or phenomenon returns to an indeterminate state until once again, the observer takes a measurement or otherwise attempts to perceive it. It's not true that the object or phenomenon simply disappears from the universe and ceases to exist. Nonetheless, the phenomenon returns to a state that we can no longer know until we attempt another observation or measurement.

Let's return to that second fundamental element of quantum mechanics—the fact that an observer is required to perform some observation or measurement or otherwise engage in perceiving some perception about some phenomenon or object. Let's consider what this means. The concept of an observer seems like a simple, straightforward concept. Like everything else we've been discussing, though, pinning down a definition may be somewhat elusive. If we really want to get our minds around quantum mechanics, then it seems terribly important to pin down a really good, reliable definition of "observer."

Unfortunately what we discover in quantum mechanics, and not at all unlike the situation in relativity, is that the concept of the observer is a bit vague. I would even go so far as to say from my experiences that scientists as a group possess and employ a range of definitions that vary broadly by individual, field of specialty, and application. How could this be? If the concept of the observer is crucial to the fields of quantum mechanics, particle physics, cosmology, relativity, and astrophysics, I would think that a common, uniform, tightly defined, universal definition of "observer" ought to exist.

I would further think that the absence of such a definition might allow inconsistencies to develop when trying to bridge these disparate fields of science. When I've spoken to scientists in these different fields, it's clear that a common, concise, universally accepted definition is either not available, not used, or not even considered important. I've been provided answers as varied as "it's self-apparent" to "an electron could serve as an observer." I wasn't convinced that everyone was on the same page, to be honest.

The absence of a common, concise, universally and scientifically accepted definition of "observer" seems like a really big, really fundamental problem to me. It turns out that there are some really good reasons the definition has eluded scientists. I'll share those with you in the following section.

I realize that the reader may be feeling a little frustrated at this point, but just hang in there. Each time I bring up an aspect of science to discuss or consider, we seem to be taking steps backward, moving us further from the goal of acquiring new insights and understanding about the universe. This process of questioning the elements of our science, even some of the sacrosanct, dearly embraced fundamental elements of our science, is crucial to our understanding of how science has produced the paradoxes in our universe that so many of us have learned to shrug off and live with. I assure you that there is a path which results in the elimination of paradoxes. Just try to keep this concept in the front of your mind as we proceed on our journey throughout this book: These paradoxes don't exist in the universe; they only exist in our minds.

Paradoxes arise as a result of incomplete or flawed assumptions and/or understandings. If we can eliminate the incomplete and flawed elements in our science and in our understanding, then we stand an excellent chance of eliminating the paradoxes. And, the most effective way I know of doing this is to return to the basics and question away. We need to question each element of our understanding of the universe, fill in the gaps, and scour away the flaws as best we can. We do this with the benefits of hindsight and a growing, precise database of scientific observations.

In order to emerge with an improved understanding of our universe, we're going to have to take some significant steps backward, unlearn a bunch of things we've each held as true since as long ago as grade school, and re-examine the ways in which we view our world. When we complete these exercises, we'll return to the double-slit experiment and see it in a whole new light. We'll see that the double-slit experiment actually provides us with the definition of "observer" that I believe scientists have been seeking, whether or not they recognize the need for it.

Get ready: We're going to take a few more steps backward during our questioning process before we begin to take our first bold steps forward in a new direction.

Chapter 3

Perception of Perceptions

Let's think about the process of observation. How do we observe? Humans possess a variety of sensory organs and capabilities. We can feel reasonably confident that our senses are somehow involved in the process of observing in some very key way, but that our senses are only part of the larger process of perception and understanding. To investigate this question, let's turn our attention to an age-old, somewhat humorous, cliché of a question: If a tree falls in the forest, and no one is around to hear it, does it make a sound? We're going to tear this question into little pieces before we're done with it, but allow me to proceed directly to the one and only possible correct answer: no, it doesn't. Not at all.

At this point, I expect that you're assembling arguments in your mind as to whether this answer is correct, so please be patient as I explain that there really is only one correct answer. You'll soon understand why the answer is "no" and why this question is absolutely crucial to eliminating the paradoxes that "exist in our universe," as we fondly and inappropriately say.

Let's examine the question in a tad more detail. I interpret the question to mean that a tree falls to the ground. That seems fair enough. I can accept that premise. The tree strikes the ground—of that I'm also completely confident. The question states that there's no observer, that there is absolutely no one around. That's an assertion I'm willing to accept. In fact, up until this point, I have absolutely no qualms with the question.

Now let's move to the last piece of the question, namely: "Does the tree make a sound when it strikes the ground?" I hope you accept my re-interpretation of the question and agree that I haven't altered its original intent.

This last part of the question is a trick! The tree surely smashes into the ground and, in so doing, it moves the ground and the air that surrounds it. There's a great deal of movement in the form of vibrations over some portion of the acoustic spectrum, but there's no sound. Sound doesn't exist in the vicinity of the fallen tree, or anywhere else in the universe, other than in the mind of an observer. Not coincidentally, this is also the only place where the universe's paradoxes reside.

I want to emphasize that this is not an exercise in semantics whatsoever. If we are truly motivated to understand a universe without paradox, then we need to be excruciatingly precise in the questions we ask, the definitions we form, and the conclusions we draw. Of course, we use language as the means to frame these questions and ponder the responses. All languages have the potential to introduce misunderstandings and ambiguity, and the English language is particularly notorious for this. As we proceed through the book, we will develop a very short glossary of fewer than twenty words. We will minimize any confusion or ambiguity in the use of these terms throughout the rest of the book.

So, please take my word for it: I'm not taking you on some sort of wild goose word chase; it's crucial to distinguish between the concepts of vibration and sound. Our falling tree generates lots of vibrations, but no sound at all.

Let's examine the relationship between vibration and sound for a moment. Without acoustic vibrations in our surroundings, there would be no sounds to perceive. Acoustic vibrations are necessary but, alone, they are insufficient to produce sound. To produce sound, we need at least two other fundamental ingredients. We need a sensor to convert acoustic vibrations into electrical or electrochemical impulses. For us, that would be our ears. The final item is a processor which can interpret those impulses and create the perception we experience. That would be our central nervous system, including our brain.

Let's say we possess advance knowledge that a particular tree in a particular forest is about to fall to the ground. Now, let's place some human observer near that tree, at a safe distance, so he doesn't get crushed (since this example involves sound and not touch, which we will discuss, too, but later). Now let's allow the events to unfold in slow motion. That tree proceeds to fall and it smacks into the ground with enormous force. The ground compresses under the force of the fallen tree and begins to undulate, setting into motion a cascading series of vibrations within the ground that are transferred to the air. These vibrations emanate out quickly from the area of the fallen tree and travel to the ears of our observer. The vibrations are then transferred to the ears' tympanic membranes, which begin to resonate. These motions are transferred through the bones and apparatus of the ear until they reach the cochlea. The cochlea then translates these vibrations into electrochemical impulses, which travel along nerves of the central nervous system until the signals reach the auditory processing center of the brain. This part of the brain then interprets the incoming electrochemical signals into a form of information that is useful to us. This information is transferred to our consciousness where, finally, we perceive sound.

Notice the words I chose: *where, finally, we perceive sound.* I avoided the alternate phrase, "where we hear sound," because that's imprecise and just an illusion. We don't hear sound; after all, we hear vibrations, which we perceive as sound.

This is a strange concept to many people, so let's take a moment for it to settle in. There are a few degrees of separation, as you now see, between vibration and sound. We don't directly perceive the world around us—ever. The one and only thing we can ever actually perceive is our perceptions. And the only place where these perceptions exist is in our minds. The perceptions don't exist outside us in the general universe. Rather, they exist only within us within our inner universe, so to speak. And just like there's a unique relationship between vibrations and sound, so there is also a unique relationship between the outer universe and our inner selves. This, of course, is completely unavoidable.

I'd like to try to vary this example just a little bit to help drive this important point home. The example is going to be a little unrealistic based on our everyday experiences, but let's go for it anyway. Let's establish the range of human hearing as acoustic vibrations operating within a frequency range of 20 Hz to 20,000 Hz. (Hz is a unit we use to mean vibrations per second.) So 20 Hz is associated with a low-pitched, deep-rumbling sound while 20,000 Hz is associated with a high-pitched, shrill sound. Human hearing range varies quite a bit from person to person, but this range is reasonable for our example. Now, let's say, hypothetically, that the falling tree only produces vibrations outside of this range, perhaps only in the range of 25,000–30,000 Hz. Now let's play out the new falling tree scenario.

The tree smacks the ground, just as before. And just like before, the ground trembles and the air vibrates. Now the vibrations reach the observer's ears and the ears' auditory sensing and conversion apparatus don't even notice that vibrations have reached them. Our ears simply don't operate in this range. As a result, no electrochemical impulses are produced and the brain just goes about its other business, and no sound is perceived by the observer.

Now let me ask you: Did this particular tree make a sound even with an observer present? It's easy in this case to truly appreciate the differences between vibration and sound. Here's the kicker: Vibrations are produced, all around us and other observers, all of the time that observers don't perceive because their sensory apparatus just doesn't operate in all possible regions of the spectrum, regardless of whether it's the acoustic spectrum or the electromagnetic spectrum. To be sure, there are other creatures that can sense vibrations in the 25,000–30,000 Hz range we assigned to the falling tree; if they could talk, they would probably have exclaimed, "What was that?"

I'd like to cover a couple more examples before we attempt the first entries in our glossary.

Let's shift to the sense of sight for the next example. We know today that light is one form of electromagnetic energy, thanks to Maxwell and others who figured this out in the latter half of the 1800s. It's not unusual for us to split the electromagnetic spectrum into two very unequal pieces: visible light and everything else. By far, the biggest part of the spectrum is the piece we refer to as "everything else." In this part of the spectrum we find radio waves, microwaves, infrared radiation, ultraviolet radiation, x-rays, and gamma rays.

As shown in Figure 4, the electromagnetic spectrum spans an almost inconceivable range of frequencies beginning with radio waves at a frequency of about 10^4 Hz ending with gamma rays at a frequency of about 10^{20} Hz. (Of course, the spectrum continues beyond the range I'm describing here, but I'm focusing on just the portion of the electromagnetic spectrum from 10^4 to 10^{20} Hz.) Within this excerpted spectrum, visible light represents only a tiny sliver of frequencies between 400 and 800 × 10^{12} Hz (or terahertz [THz]). Our sensory apparatus, our eyes, are truly limited in their ability to sense different portions of the electromagnetic spectrum, so we devised technology to permit us to "see" in these other portions of the spectrum.

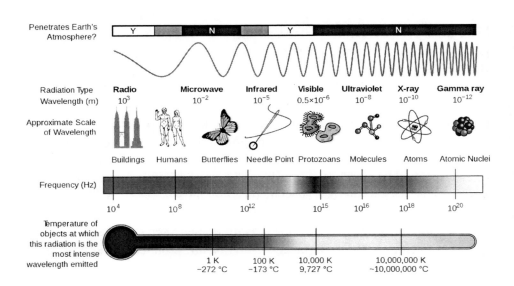

Figure 4. A diagram portraying the electromagnetic spectrum, showing the type, wavelength (with comparative everyday examples), frequency, and associated black body emission temperature. Source: Wikipedia.

Once again, I placed the word "see" in quotes since, as we all now know, we don't see light. Waves of electromagnetic energy reach our sensory apparatus—our eyes—in which photons (carriers of the electromagnetic force) get absorbed by special cells that line the retinas of our eyes (rods and cones). The absorption of the various frequencies of photons in our eyes produce corresponding electrochemical signals that are then sent to the vision center of our brain. They are interpreted and converted into a form of information that is then sent to our consciousness and perceived as images.

As for me, I personally have no difficulty in understanding the difference between electromagnetic energy and vision, or between light and vision. You see, I'm color-blind and reminded all too often that my perceptions don't match those of "normal" sighted persons. The perceptions of a color-blind individual are limited with respect to those of a normal sighted person. Almost without exception, whenever people learn of my color blindness, they ask one of two types of questions, or both. They point to some article of clothing that they're wearing and they ask me to identify the color, apparently so they can satisfy themselves that I'm really, truly color-blind and not simply pretending to be, which of course doesn't make any sense any way you "look" at it.

The other sort of question goes like this: "So, what does _____ [fill in a color of your own choosing in the blank] look like to you?" Although it's a perfectly natural and innocent question to ask, after a short moment of reflection, it might occur to you just how absurd the question really is. You see, I've only ever had one set of optical sensory apparatus with which to perceive electromagnetic energy and it doesn't allow me to see what normal sighted people see. While I was a young child growing up, my parents faithfully and consistently pointed to a particular color and they always called it the same color each time—at least that's what I've chosen to think. So, every time I'm shown a patch of the color red of a particular hue and saturation, it always looks the same to me. But how in the world could I ever hope to explain my perception of that color in words in any meaningful way? It simply can't be done. The closest I can come to verbalize an explanation of my type of color blindness is this: Imagine turning down the volume of the red component of any color and that's close to what I must perceive. You might have noticed that I used a term normally associated with sound (volume) to get my point across.

After more thought, perhaps that's not so strange since, as you might recall, we only perceive perceptions, so analogies that mix and match different senses is utterly and innately understandable to my questioners. In any event, once you tone down the red component, brown looks like green, purple looks like blue, and pink looks like gray. To me, red is a very dull color and I always wondered why warning signs were intentionally made to be so obscure. Blue is another story to me. Blue just screams at me! Thank goodness for blue.

It's my "faulty" color perception that actually got me thinking about perceptions of our world. In my world, there's absolutely no way to avoid it.

So now let me ask some questions of the reader: Do all of you normal-color-sighted people out there "see" precisely the same color information? What does red look like to you? Or, green? Or orange?

Let those questions sink in for a moment....

Okay, so let's consider the answer, which is: There is no sure-fire way of knowing. Each of us has developed our own senses of perception within our own selves using the only sensory apparatus we happen to possess. And although we create a common sense of color perception through the mixing and matching of colors considered to be either pleasant or unpleasant, we can never actually examine whether anyone's perceptions are identical to those of another. Through language and pictures we communicate our perceptions of color and thereby attempt to harmonize our perceptions of color with those of others.

Let's look at another example to understand vision. This example involves books, just like this one. Open a book, either on your eBook reader or the old-fashioned kind with bound paper sheets, and what do you see? In fact, whether you are reading this book or any book, what are you seeing?

The quick but incorrect response is that the book is full of words and reading the book means reading the words printed in it. But, that's not actually the case. The only thing printed on the pages of the book is ink in a color that contrasts with the color of the sheet of paper. There are only two elements of the page: those parts of the page on which there is no ink and those parts of the page on which ink has been printed. The printer didn't print words or even letters, although that's the illusion of perception. Rather, the printer printed ink in some places that contrast with the parts of the page in which no ink was placed. When you read the book, your brain interprets the patterns of ink into letters and words which you read effortlessly. In fact, before you even begin to perceive words, your brain first has to interpret the image sent to it by your eyes to determine where on the page to begin scanning for patterns that it will ultimately interpret into words.

Let's take it once again step by step. You settle in to your comfy reading chair, having already made certain that you turned on a reading lamp to ensure that photons of visible light will flood the page you choose to read. Photons reflect off the regions with ink and without ink; the resulting light is focused onto your retinas by your eyes' lenses. This pattern consists of a duplicated image of the page but still consists of only a pattern of illuminated and less-illuminated spots each representing either ink or the inkless paper background. Each rod and cone in your retina produces a signal corresponding to the more or less illuminated regions of the inked pattern. These stimuli produce electrochemical signals that are sent through the optic nerve to the vision center of your brain.

Your brain interprets the individual signals and assembles the signals into an image comprised of letters and words, which then gets passed to your consciousness, as well as the portion of the brain that then further interprets the letters and words into concepts. The concept embodied by each word is combined with the other words in the sentence to create a larger concept from which we perceive context. These individual concepts and contexts ultimately allow the story that the author is trying to share with you to blossom into a colorful account. This duplicates the concepts that began as simple electrochemical signals in the brain of the author when he wrote his story.

Our brains don't always construct words so quickly from the pattern of ink and non-ink that we're observing. If that weren't the case, everyone who plays Boggle or Scrabble would compete at the Grand Master level all the time. What makes word games like these and others so entertaining is the fact that our brains don't reliably and easily assemble the patterns sitting right in front of our noses into words that we can actually perceive. Game developers count on this phenomenon to create entertaining games and exercises. How many times have you said that something was right in front of your face and you didn't see it? It may be frustrating at times, but it's the nature of the process of perceiving perceptions.

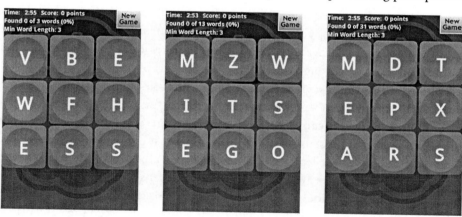

Figure 5. Screen shots from the WordUp! game in order of increasing difficulty: 3 words (left), 13 words (middle), 31 words (right). Answers in Appendix IV. Courtesy: Anthrological, Inc.

If you want to have some fun and understand the difference between there being something to observe and actually observing it, then try your hand at a word game called "WordUp!" by Anthrological that I often play on my Android-powered phone. This game reminds me of just how little we observe phenomena even when they're staring us in the face. Set the game on a 3 × 3 board, which is comprised of only nine measly letters, and see how difficult it is to find the words hidden within that small matrix of letters. It's not like the letters are hidden. The letters are all in plain view, yet detecting all of the words they make can be ridiculously difficult. Construct words by connecting letters on adjacent tiles, horizontally, vertically, or diagonally. Go ahead and try each of the three puzzles in Figure 5. The puzzles increase in difficulty from left to right.

So let's wrap up our discussion of vision with a little quiz: If a star in the sky goes nova and there's no one to observe it, does it produce light? I do hope that we agree on the following answer: No, it certainly does not! It emanates electromagnetic waves but, without an observer with the proper sensory equipment operating in the right portions of the spectrum and the ability to convert the electromagetic waves into a perception, there is no light; light is the result of perception. No observer means no perception, which means no light.

We're going to look at one last example to drive the point home before we start on our glossary, after which we'll deal with the paradox of the double-slit experiment. This next example involves the sense of touch.

Our sense of touch presents yet another slippery slope in our quest to observe the world around us. Like each of the other senses with which we probe our surroundings, we can only *perceive* the perception of touch. And just like our other senses, the perception of touch begins with a stimulus, this time an object or phenomenon interacting with the nerve endings in our skin. As something presses against our skin, special nerve endings convert the interaction into electrochemical impulses that travel through other nerves to the part of our brain that processes and interprets the impulses into a sensation that our consciousness can understand and use. Once again, we only perceive the resulting perception.

The sense of touch can be particularly deceiving, especially given the various manifestations that the sense of touch takes. The sense of touch involves pressure, pain, hot, and cold—at least that is what we typically think, since that's what our kindergarten teachers tell us. In actuality, our senses of hot and cold are not what we normally consider them to be. I'll make this side discussion brief so we don't get too off topic, and it does relate to the concept of perceiving perceptions.

We actually don't sense hot and cold, or more accurately, we don't perceive the perceptions of hot and cold, although it's quite common for us to interpret our perception one step farther and decide whether an object is hot or cold. When it comes to temperature-related sensations, our receptors are actually quite limited. They can only sense the direction of heat flow; they can't actually sense a temperature.

This statement requires a little clarification, if this is the first time you're encountering this concept. Let me start by asking a question of you. If you have two objects, say a steel butter knife and your napkin at the dinner table, sitting near one another in the same room at the same temperature, did you ever notice and wonder why the knife felt cold and the napkin didn't? Well it's because of the fact that we sense the direction and speed of heat flow into or out of our fingertips. The knife feels cold because metal is a good thermal conductor (a good conductor of heat). The moment your skin touches the knife, the heat from your finger transfers into the metal and is then conducted outward and throughout the metal in the knife. Thus, the metal knife "feels" cold.

The situation is different with a napkin. Fabric is a poor thermal conductor. In fact, our clothes are produced of fabric for that reason. Well, that and the fact that they're also not transparent. To be sure, some fabrics are better thermal insulators than others, but most are exceedingly poor thermal conductors. So, what happens when we touch our napkin? For the first moment after our skin touches our napkin, a small amount of heat flows from our skin to the fabric napkin. The fabric is a poor thermal conductor, so the heat has no place to go. So, once we touch any spot on the napkin, a small amount of heat from our skin heats up the spot very quickly and the heat flow out of our skin stops. Like every other one of our senses, there's a small time delay that is usually not detectable to us, and the heat transfer process involving the napkin happens on a timescale similar to this delay. So it's very unlikely that we'll ever notice the momentary "cold" when first touching the napkin.

If we touch something warm—higher than our body temperature—then heat from the object flows into our skin, we perceive the perception of heat flowing into our skin, and we decide the object is warm. If we touch something very hot or very cold, then the act of touching the object results in greater and faster heat flow. We perceive the corresponding perception, and the result is that we have "sensed" the object's hotness or coldness.

Our body also has special built-in features called reflexes that tighten muscles to pull our hands or other body parts away from an object if the heat flow exceeds certain thresholds. In fact, these reflexes act way faster than the process involving the perception of temperature, since reflexes involve a shorter nerve path and one that bypasses the touch processing center of the brain. Only after a reflex action has occurred do we get the delayed sensation of hot, cold, or pain.

So, let's get back to the final example of touch I promised you. First, like each of the other examples, we need to take some important steps backward to break the process of touch down into simple, understandable, well-defined elements. Here are a few questions to consider:

◊ What is touch?

◊ At what point have I actually touched something?

◊ What is the nature of the boundary between what is me and what I touch?

◊ What am I actually feeling?

I hope these questions don't make you feel that touch is any more complicated than our other senses; it's just that it's really important to break the perception of touch down to its basic elements, to its first principles. Within our discussions so far, we talked about the perception of light and sound. These senses involve phenomena reaching us from a distance. But touch is different in that touch can only occur close to us, within our reach. Touch is, therefore, a far more local and intimate experience. And whereas two people can generally feel that they are sharing an identical auditory or visual experience with their friend, it's just not possible for two people to be touching and experiencing the same patch or portion of an object at the same moment. This aspect of touch allows us to begin investigating and considering the uniqueness of each observer's perceptions and perspectives. These elements will be useful as we construct our glossary.

To investigate the nature of touch, we need to leap to a discussion of atoms. In fact, throughout the book, we will often need to return to discussions of atoms. We're going to start out with a discussion of the atom based on the concept of the Bohr atom that will in turn lead us to the concept of the Bohr radius of an atom.

Niels Bohr was among the scientific giants that helped to usher in the Quantum Age in the early part of the twentieth century. This is a relatively modern perspective of the atom given that the concept was first devised by pre-Socratic Greek philosophers nearly twenty-five centuries earlier. Throughout our discussions of the atom, we will certainly cover the earliest concepts of the atom, not only because they are utterly fascinating (well, at least to me), but also because the ancient founders of the atomistic concept considered important elements of the vacuum that we need to include in our path to establish a paradox-free framework for understanding the nature of the universe.

As many of you are probably already aware, an atom is the basic, chemically indivisible unit of an element. (Each of the elements is listed in the periodic table.) Any atom consists of a nucleus and some number of electrons orbiting the nucleus at a distance. The nucleus consists of one or more protons; furthermore, the nucleus may contain some number of neutrons. The number of protons in the nucleus defines the atom's basic identity. Every atom of a particular element possesses the same number of protons. And the number of protons the atom possesses is called its atomic number. Each element therefore possesses a unique atomic number.

Different electrical charge states are associated with each of these subatomic particles. Protons each possess a single positive charge, while each electron possesses a single negative charge. Neutrons, on the other hand, don't possess any electrical charge. Neutrons therefore possess a neutral electrical charge state, as you might have presumed from its name. The number of protons and electrons within an atom are identical, so the atom itself has a neutral electrical charge state.

Certain atoms of a given atomic number may possess different numbers of neutrons, which result in differing atomic weights. An atom's atomic weight is simply the sum of the numbers of its protons and its neutrons. The various atomic weights associated with a given element (or atomic number) are each referred to as isotopes of the element. The abundance in nature of each of these isotopes varies, so the atomic weight assigned to each element in the periodic table represents a geometrically weighted average of the isotopes that takes the abundance of each individual isotope into account.

Now, let's consider what happens when we attempt to bring together these different particles. As you are also probably aware, "like" electrical charges repel each other and opposite electrical charges attract each other. In other words, if we attempt to bring together two electrons, a repulsive electromagnetic force carried by photons will attempt to drive the electrons apart from one another. The same is true of protons: If we attempt to bring together two protons, a repulsive electromagnetic force, again, carried by photons will attempt to drive them apart. Photons are the carriers of the electromagnetic force and they are the only way the electromagnetic force is conveyed.

Well, then, you might ask, "How is it that protons live together in peace and harmony and in close proximity to each other in the nucleus?" We are already familiar with the electromagnetic force, but to answer this question, we need to introduce another fundamental force: the strong nuclear force. As we pry two protons closer and closer to each other in defiance of the electromagnetic force, we eventually get the protons close enough to each other that the strong nuclear force becomes dominant and the two protons become and remain close buddies. This energetic process happens all the time inside stars as two individual protons (hydrogen nuclei) are fused together to form a single atom of helium with a nucleus containing two protons, thereby providing helium with an atomic number of two. This process is called fusion.

The situation with neutrons is somewhat similar. Neutrons have no electrical charge associated with them, so they don't really give a hoot about being near another neutron. From the perspective of the electromagnetic force, neutrons are just plain apathetic. Yet, once again, when they get close enough, like in an atom's nucleus, the weak nuclear force becomes dominant and holds the various neutrons together harmoniously within its happy nuclear family.

Let's consider the atomic situation. We notice that three fundamental forces are at work within the atom: the strong nuclear force, the weak nuclear force, and the electromagnetic force. Within the atom's nucleus, the strong and weak nuclear forces keep the protons and neutrons tightly bound. The positively charged nucleus comprised of protons and neutrons is surrounded at a distance by a bunch of electrons flying this way and that.

These electrons constitute a shell of negative electrical charges that shield the positive charge of the nucleus buried deep within it, so that, at some distance outside of this shell and away from the atom, the atom's negative and positive electrical charges cancel each other out to yield an electrically neutral atom. The radius of this electron shell is referred to as the atom's Bohr radius. The Bohr radius of atoms is on the order of an angstrom, depending upon the particular element you consider.

(For the sake of falling prey to the conventional comparison regarding such things, a human hair is about one million angstroms thick. This comparison is utterly ridiculous, though, since there's no standard for the width of a human hair. Depending upon the person from whom you pluck the hair, the value of its width could vary enormously. Nonetheless, you get the point, I'm sure, that the width of a human hair is small and the size of an angstrom is really, really tiny in the scheme of our everyday experiences.)

Generally speaking, the Bohr radius increases with atomic number, although the specific configuration of the electrons in their shells also plays an important role, but not one with which we need to concern ourselves.

Let's recap again. In the nucleus, where protons and neutrons reside, strong and weak nuclear forces dominate. As we move out of the nucleus and then into the electron shells, the impact of the strong and weak forces drops quickly and the electromagnetic force dominates. Finally, as we emerge from the electron shells and move farther away from the atom outside of its Bohr radius, we reach a point at which the influences of the positive and negative electric fields of the nucleus and electrons cancel each other out, and we achieve a net zero electric field.

With these concepts in mind, let's now get back to the topic of touch. Okay, look around for something to touch. I'm sitting at a desk writing, so my obvious suggestion is to consider reaching out to touch a desk or other piece of furniture in your vicinity. As our fingertips approach the furniture (or anything else for that matter), we begin to play the previous scenario in reverse. Specifically, we're moving closer and closer to the atoms of the desk.

In this example, the skin at the surface of our fingertips is what we normally consider to be the boundary between us and anything external to us. Let's just go with that for now. So, our fingertip is moving through the empty space between our fingertip's edge and the surface of the desk. At these distances, the atoms in our fingertips look electrically neutral to the atoms of the furniture and the atoms of the furniture look neutral to the atoms of our fingertip.

We move our fingertip ever closer to the furniture and close the distance separating them. Now, the situation has changed. We've now entered the regime at which electromagnetic force dominates. The electric field of the electron shells dominates the situation for reasons that now are apparent. We're closer to the electrons than the proton-containing nucleus, and the shells of negative electric charge shield the effect of the proton's positive charges. So, at the moment at which we perceive that we have touched the furniture, we have, in fact, done no such thing.

The electron shells of the atoms on the surface of the furniture and the electron shells of the atoms on the surface of our fingertips have come into close proximity to each other. Since like electric charges repel, these electron shells repel each other. That "push back" represents a stimulus, initiating the creation of an electrochemical impulse that travels through our nerves to the portion of our brains that processes electrochemical impulses coming from the touch receptors (sensors) in our fingertips, which then processes those signals into something that our consciousness can perceive. When all is said and done, we believe that we have touched the furniture even though we haven't. I know you may be getting tired of hearing me say this, but all we can perceive are perceptions.

Let's recap our discussion of the perception of touch. Touch is the perception resulting from electrochemical impulses generated in special receptors that respond to the proximity of negative electrical charge arising from close proximity between the electron shells of atoms on the surface of our skin and the electron shells of atoms in the object we believe we are touching.

In each of these examples and in countless others you might choose to consider, perception of the world around us begins with a stimulus in which an electromagnetic force from the external world interacts with special receptors (sensors) in our body (self) to produce an electrochemical signal that travels through our nervous system to a specialized part of our brain that processes the signal into a meaningful form that our consciousness can recognize, and the processed signal (perception) is transmitted to our consciousness (another part of our brain). And, lo and behold, we perceive the perception and become aware of some aspect of our external world, which becomes our observation.

Sound isn't sound until the final step in a long process. Vision isn't vision until the final step in a long process. Touch isn't touch until the final step in a long process. And, of course, the same is true for smell and taste.

Let's word this another way. The stimulus begins in the world external to the self. Next, electrochemical impulses are produced by a special sensory apparatus (sensor) and transmitted within our body (self) to a part of our brain that processes the signals into something useful. A new set of electrochemical signals (perception) reaches another part of our brain in which we interpret the signals as sound or sight or touch or smell or taste (perception of the perception). And, finally, since the mind is housed within the collective set of structures called our brain, the sound/sight/touch/smell/taste exists nowhere else in the world but within our minds, which might feel a bit numb at this point.

Remember that the stimuli that come at us from the external world must first be capable of interacting with our sensory organs or a perception can't possibly result from the process. In other words, acoustic vibrations caused by the electromagnetic interaction of atoms with one another that reach our ears must be within a certain part of the acoustic spectrum to result in a perception of sound just as electromagnetic energy must be within a certain part of the spectrum to result in a perception of sight. If a chemical vapor doesn't interact with our olfactory organs in just the right way, there is no perception of smell and the vapor is considered odorless by the particular observer. The same is true for substances that are presented to the taste buds on our tongues. The absence of just the right interaction with the substance we place into our mouths results in a null sense of taste or "tastelessness."

To a different observer with a different set of sensory organs that possess different capabilities, what is observable to us may be unobservable to that particular observer. Conversely, what is potentially observable to some other observer may be unobservable to us. I think you'd now agree that huge swaths of our universe may be unobservable to a particular observer possessing a particular set of sensory apparatus occupying a particular perspective.

Now, we're getting somewhere. I hope you agree.

Science has absolutely no way of knowing the ways in which the perceptions of one individual compare and contrast with those of another individual, generally speaking. In other cases, the differences are quite clear. Many unfortunate individuals have an impaired sensory apparatus, like me. I completely understand that a non–color-blind person "sees" something quite different than I do when observing the landscape or a sunset or a movie. Other individuals might possess impairments associated with other parts of the perception process. In fact, there are people who "hear" colors and others yet who "see" color-coded letters or numbers when they look at a sheet of characters printed only in black ink on white paper. In fact, just about every variation you might consider has been observed and studied by medical researchers and psychologists.

So, it's not just beauty that's in the eye or mind of the beholder: Every perception we possess exists exclusively in our minds and nowhere else in the universe.

It seems like a good time to mention again that the only place where paradoxes exist is also exclusively in our minds. This is not a coincidence, of course, since the universe is not really paradoxical. Paradoxes arise from faulty or incomplete interpretations and those interpretations can exist in only one place in our vast universe. That one place would be in our minds.

So, allow me to introduce the first entries in our glossary:

◊ **Perception** – the collection of information by the self about the internal self (internal states) or about the external universe (external states) through the capabilities of its unique sensory apparatus.

◊ **Consciousness** – the act of synthesizing information about the internal self with information about the external universe.

◊ **Self-Awareness** – the sense of self unique to each self.

Please spend a few moments familiarizing yourself with these terms and the definitions I've proposed. These terms represent key concepts that will prove crucial in our quest to eliminate paradoxes.

Our glossary is started. The remaining terms will be introduced in subsequent portions of the book.

I want to engage in one last departure before we do away with that pesky double-slit paradox and delve into other scientific mysteries.

We've done a good job of breaking the process of perception down into its components. I'm confident that by now the reader has become at least somewhat weary of taking steps backward in order to take important leaps forward, even if the reader understands and appreciates the need to do so. But we need to talk about a related topic before we can agree that we've established a firm, common foundation from which we can propel ourselves into further analysis and understanding of our universe and the paradoxes that modern science produces.

Let me just blurt it out and get it over with. *Virtually everything you've ever learned about anything is wrong!*

Yeah, I realize it's a strong statement but now that I've dug my hole, I'm going to try to defend it and convince you that it's true—especially when it comes to science and its most useful tool, math. I really hope the experiences that I'm about to share with you resonate in some way.

I realize that not everyone out there was as focused on science and math as a young student as I was, and for the most part, the stories I'm sharing with you relate to science and math. However, I do believe it's possible that those of you who focused on history or English or foreign languages or just about any other grade school subject had similar experiences.

The experience goes like this. One year in school, say seventh grade, your science teacher fills your head with lessons about a science topic (like physics, for example, and you spend the school year conscientiously learning the material). Then, a year or two later, as an older, more capable, more mature student in the ninth grade, you take another year of science that covers physics. Imagine that the first thing your teacher tells you is that everything you had previously learned about physics was wrong!

After the initial shock subsides and you proceed with your studies, you learn that what the teacher really meant was that the last time your teachers presented material to you, the curriculum incorporated certain age-appropriate simplifications that enabled you to grasp the material.

Well, that process continues in many ways throughout our school careers and, in reality, extends past school and into our adult lives. There is simply so much information and so many perspectives on any subject—especially in science, in my opinion, since I'm most comfortable saying this about science—that simplifications are commonplace.

We must always produce simplified models in our minds in which we incorporate the elements we consider or interpret to be the most important. We use these simplified models because that's really the best we can do as humans. For most practical purposes, these simplified models are adequate for our daily lives, but we must never forget that there is a difference between the simplified model we use and reality, especially when we encounter a paradox. We shrug off many of life's little paradoxes on a daily basis because we need to constantly turn our attentions to the onslaught of stimuli and media that characterize human society in the early twenty-first century.

Other elements of the models we use and the curriculum we've been taught as students, however, are just plain wrong in the old-fashioned sense of wrong. Let's consider a few short examples.

I really enjoy picking on geometry when it comes to using examples to demonstrate this particular point. Once again, it's no coincidence that I'm choosing geometry to press my point. First of all, geometry is an easy target and one for which many of us long for some sort of payback. But, more importantly, geometry is the set of tools that we commonly use to comprehend and discuss the shape of our universe; this will become crucially important in the chapters in which we discuss such topics as the Big Bang (and its remnants), as well as the related topic of our apparently expanding universe, culminating in our discussions of dark matter and dark energy.

I'll try to be true to my word and make this discussion brief.

There's a concept called the Euclidean plane upon which Euclidean geometry is based. Euclid of Alexandria, was a Greek mathematician, often referred to as the "Father of Geometry." A Euclidean plane is flat and extends infinitely in all directions. We can apply a Cartesian coordinate system to this plane in which the x-axis is at right angles (orthogonal) to the y-axis. At each and every point on this plane, then, lines parallel to either the x-axis or the y-axis are always orthogonal to each other. Hopefully, I haven't lost you. Just hang in there with me.

I took geometry in eighth grade, and I'm confident that other students who took geometry were taught something like the following: Parallel lines on a Euclidean plane meet only at infinity. Now let's break this down so we can better understand the situation.

The concept of infinity is central to the situation at hand. The basic elements of the situation, then, are that we have a perfectly flat, infinite plane upon which we have placed two perfectly straight, perfectly parallel lines that extend infinitely in each direction. Now let me ask you this: If we buy into the concept of an infinite flat plane and two infinite, parallel lines placed upon it, then where would the lines intersect? Well, they don't intersect. In fact the concept of parallel lines on a flat plane defined from the beginning is that *the lines don't intersect.* We all knew full well that we couldn't say this on an exam if we wanted a good grade, but I'm pretty sure that each of us thought it, right?

So, where's the problem?

I once took a class on statistics. The teacher wryly defined statistics at the start of the course: "the art and science of proving whatever it was that you wanted to prove in the first place." Now, before I start getting nasty messages from statisticians, I want to remind you that these were not my words. I keenly recognize that this is not the most appropriate or respectful way to introduce statistics. Nonetheless, it seems to me that there's at least a grain of truth in there. Of course, the same holds true for math, science, and just about any other field of study you can consider. The fact is that it's terribly difficult for us to separate our objectivity and our subjectivity. We'd have to be someone other than ourselves to prevent our personal biases from creeping into our endeavors. Scientists, mathematicians, statisticians, and actuaries are all human the last time I checked, and bias seems to be, at least to me, an inherent component of the human condition.

So let's return to the problem of parallel lines meeting at infinity. Where does the specified plane exist? Where do the specified lines exist? Where do these lines meet or not meet? Where does the concept of infinity reside? The answer to each of these questions is the same. These items exist only as concepts within our minds.

Infinite lines can only truly intersect if we choose to mix and match concepts of finite and infinite. As humans, we just can't get away from the fact that when we attempt to conceptualize an infinite plane, the best we can really do is to visualize an image of a small portion of our mind's infinite plane. Within this framework, sure, I can then accept that parallel lines meet at infinity, especially when my everyday perceptions can easily deceive me. After all, if I stand in the middle of a long road and look out along it, it would certainly appear as though the road were converging to a point in the distance. The funny thing is that, of course, the road really doesn't diminish or else we'd all be paying higher car insurance rates than we currently pay. Perception and logic combine in such a way that we are willing to accept and embrace the concept that parallel lines meet at infinity. The introduction of perceptions interferes with our abstract reasoning skills when we attempt to introduce our interpretation of that pesky and elusive concept we call reality. And, like all other concepts, the concept of reality can only exist in one place in the universe and that's (hopefully you guessed it already) in our minds.

When it comes to conceptualizing quantity, the human mind is truly limited.

In fact, each of the quantities that we can truly conceptualize and get our minds around is approximately zero in the grand scheme of things.

Let's go through a simple exercise together now.

I always found it entertaining to consider the limitation of my own mind in dealing with numerical concepts, even though I spent so much of my life involved in performing scientific and engineering calculations involving outrageously large or vanishingly small numbers applied to various abstract applications. Let's try a simple little exercise.

If I ask you to imagine a single object, then, like most people, you'll create some image in your mind of a lone object of some sort in the center of your mind's whiteboard. Perhaps you imagine a simple geometric figure, such as a dot or a sphere or a square or a tick mark or something entirely more imaginative, like what Cuddles must have looked like. Fine.

Next, imagine two objects. Most people imagine two objects placed close to each other, usually side by side. Now, imagine three objects, and then four. Most, but not all, people who perform this part of the exercise imagine three objects placed on the corners of a triangle and four objects placed on the corners of a square, but there's no right answer, after all. (I didn't pull Cuddles out of the cabinet, so this isn't a pop quiz!)

Now, continue to increment the number of objects in your mind by one at a time. By the time most people get to six, the mental image has collapsed to two rows of three objects. At a quantity of eight, the mental image described by most people collapses to two rows of four objects. And, usually by about ten or so objects, it becomes exceedingly difficult for most people, including me, to continue seeing discrete objects in our minds at all.

We begin replacing individual objects with symbols to help us continue the exercise. For instance, sixteen objects may be imagined as four somewhat vague clusters. Then you need to zoom in using your mental magnifying glass to resolve that each cluster is comprised of four discrete objects. So, the lesson here is that most minds must collapse the visualization problem down to helpful constructs that are themselves limited to small fundamental quantities, like four, or five, or six. The bottom line then is this: Fundamentally, it appears as though I can't even get my mind fully wrapped around the not-too-terribly ambitious concept of ten or twenty objects *and* I'm hoping to use this very same mind to comprehend the vastness of the universe and the richness arising from diversity of human thought? Ouch, that's a bubble-buster.

Okay, it's time to move onward and begin to slay paradoxes!

Since the discussion of the structure of the atom is still fresh in our minds, let's dissect the atomic structure one more time. This time, however, let's look at the volume that the atom occupies and break it down into its volumetric components.

For this exercise, we need to choose an element from the periodic table. (See Figure 6.) Let's examine an atom of the element carbon, since that's a particularly important element for you and me. Along with most of the living creatures on our planet, you and I are composed of a lot of carbon. We consult the periodic table and learn that carbon has an atomic number of six, meaning that the nucleus of the carbon atom possesses six protons. To be a neutral atom, we now know that the electron shells must contain an equal number of electrons. So far, we know that there are six protons and six electrons in an atom of carbon. How do we determine whether carbon possesses any neutrons?

Further inspection of the periodic table informs us that the average atomic weight for carbon is just a tad over twelve atomic mass units. Six of those atomic mass units are already accounted for—i.e., the six protons of its y. Therefore, we can take the difference between the element's atomic number and atomic weight as an estimate of the number of neutrons we might expect to find in any particular carbon atoms we encounter. As a result, we have determined that the element carbon possesses on average about six neutrons.

Determining the basic makeup of an atom of carbon was the easy part. Now, we're going to determine the volume occupied by an atom of carbon. We decide that an atom of carbon should be approximated by a sphere. If we know

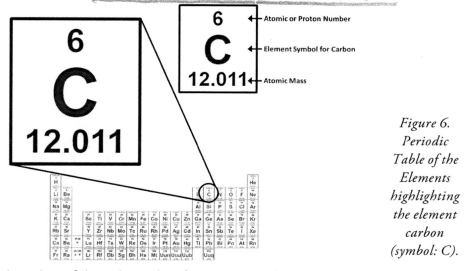

Figure 6. Periodic Table of the Elements highlighting the element carbon (symbol: C).

the radius of this sphere, then from our everyday experiences, we know that we can calculate the volume of a sphere in a rather straightforward manner by using a simple formula that some of us may remember from school. The volume of a sphere can be calculated by using the formula:

The only piece of information we need for this calculation, then, is the radius of the atom. The challenge arises when we realize that unlike volleyballs or planets or other firm, somewhat idealized spheres that we can measure using

$$Volume\ of\ a\ sphere = \frac{4}{3}\,\pi\,r^3$$

one sort of yardstick or another, the radius of an atom is elusive. In fact, this is typical of situations in the quantum world of exceedingly small phenomena.

Recall that the outermost component of any atom is its electron shells. Perhaps if we delve a little into the nature of the electron shells then we can find some straightforward way to calculate a radius.

The good news is that there is indeed structure that quantum mechanics reveals about the electron shells. The electron shells are organized into structures referred to as orbitals and suborbitals. And within these orbitals and suborbitals, electrons are flitting around in completely berserk ways that are largely unpredictable.

This is where we encounter one of our first quantum frustrations. A scientist named Werner Heisenberg summed up the problem for us. It's just not possible to know in a precise fashion both the position and the momentum of an electron at some given moment. You see, any attempt to observe the electron alters its position or the direction in which it was moving in an unpredictable way. We can determine where it was at some moment. What we do to determine where it was then alters where it's going. So, we can know where it was at some moment with precision but, in doing so, we lose the ability to precisely determine where it is in the next moment. Conversely, we can determine the direction and speed in which it was traveling at some moment but, in so doing, we lose the possibility of knowing precisely where it was at the same moment.

The more precisely we know one piece of information, the less precisely we know the other. This concept has become known as the Heisenberg Uncertainty Principle and it is truly one of the fundamental pillars of quantum mechanics. Because of the vanishingly small nature of the quantum mechanical world, this principle represents a law that can never be broken no matter how clever we become or how hard we try.

Perhaps you feel like we ought to throw our hands up in the air in exasperation and give up any hope of calculating the radius of an atom? I hope not, since the situation is far from hopeless. In fact, we end up with a couple of ways to achieve our goal.

First, although we know that the electrons are not calmly sitting in their chairs waiting for us to measure their positions before returning to the frenetic dance around the nucleus, quantum mechanics does allow us to calculate the probability associated with finding an electron at some particular position around the nucleus. So, using quantum mechanical principles, we can determine a radius around the nucleus within which we have a high level of confidence of finding the atom's electrons.

Second, we know as we approach the electron shells, the negative electric charge of the electrons will begin to repulsively interact with the electric field of another approaching atom. At some point, it becomes difficult for the two atoms to approach each other more closely. Can we use these phenomena to achieve our goal of calculating the atom's radius? Absolutely! But, you are probably thinking that, if we do this, the two electron shells are not actually touching, so this isn't really the "hard" radius of the atom. Well, you'd be right, but recall a couple of important items.

There just isn't any such thing as a "hard" boundary in the quantum world. There are only probabilities. Also, bear in mind that our everyday experience of "touching" objects is just an illusion resulting from the particular steps involved in creating the perception of touch. We never actually come into contact with the item we're touching once you understand what's happening at the quantum scale. In fact, when you really think about it, if there's really no firm boundary defining an atom in an item we touch, then, conversely, there's no firm boundary defining an atom on the surface of our skin. That might correctly allow you to surmise if we extend this thinking just a bit that therefore there is no firm boundary defining the interface between us and the world around us. That pesky perception of touch fools us into thinking that there is a sharply defined edge separating what is me (the self) on one side from what is the world around me (the external universe).

Making a long story a quantum increment smaller, scientists have used the latter approach and determined that the radius of a carbon atom is about 1.7 angstroms. We can use this radius to calculate some "hard" volume for the atom. There's really no need for me to show you the results of the calculation, but feel free to calculate it yourself, if you're so inclined.

Let's calculate the volumes for each of the subatomic constituents of the carbon atom. Recall that we have six protons, six electrons, and about six neutrons.

Let me now state what will probably be abundantly obvious to you at this point. If you think calculating the radius of the atom was tough, then calculating a high confidence value of the radius of the teeny, tiny particles comprising the atom must be ridiculously tough. By the same token, if calculating the radius of a proton is tough and we have the benefit of the proton having an electric charge, then it must be ridiculously ridiculous to determine the radius of a neutron that possesses no electrical charge state. If you thought these things, you'd be completely correct.

You get the point, so I'll just cut to the chase on this one. It turns out that the radius of a proton is about 50,000 times smaller than the radius of a carbon atom. For grins, we'll use a similar volume for the neutron. However, this is very conservative, since there's no electric field that surrounds the neutron, enabling it to "bloat" its influence like the proton can. Finally, we come to the electron.

Recent research indicates that the radius of the electron is puny, even when compared to the puny size of the proton. For this exercise, we'll just neglect the tiny volumetric contribution of the electrons.

Once we plug in the numbers and do our math (I'm just kidding, I used a spreadsheet on my computer), we find out that the volume of an atom of carbon is about 500 trillion (that's 500,000,000,000,000!) times larger than the sum of the volume of its constituent subatomic particles! That's astounding, but what does it possibly mean?

As we discussed earlier, humans in general have a tough time really being able to conceptualize a quantity larger than five or six, so how in the world can a number like 500 trillion take on any meaning whatsoever?

First of all, the analysis and the resulting ratio (one part volume of stuff to 500 trillion parts of empty space) tell us that the single largest constituent of atoms (or regular matter) is empty space. By far, nothing else comes close. When you think about it a little bit more and turn this statement on its head, an interesting thought emerges: the volumetric difference between matter and completely empty space (void, vacuum—whatever you want to call it at this point is fine) is one part in 500 trillion!

Indeed, there's a very fine line between what we perceive as nothing and what we perceive as "something!" This comes as a shocking revelation to most of us who really bought into the idea that our senses and perceptions were giving us a really good picture of the world in which we live. All we perceive are our perceptions, and our perceptions are pretty far off the mark, it seems, when it comes to perceiving the solidness of objects.

Armed now with all of our newfound insights and perspectives, let's return to the double-slit experiment.

Chapter 4

The Observer Detector

Let's perform a quick review of the double-slit experiments we talked about earlier and the peculiar results they provide.

If we produce photons and allow the photons to pass through two precisely formed slits cut into a thin, opaque panel of material onto a screen placed behind it, we observe an interference pattern. If we instead produce one photon at time and allow it to pass through the slits and onto the screen, then over time we observe the very same interference pattern. And, if we interfere with the photon in any way whatsoever by trying to observe or measure which path it takes through the two slits, our interference pattern disappears altogether.

Furthermore, when scientists perform this experiment, or any of the countless variants of it, they always observe the same results. New researchers always feel the need to see the paradox for themselves and other researchers feel like they can devise some clever variant that will overcome or sidestep the paradox. The results seem to defy interpretation but, fortunately, the results are completely consistent.

For the moment, let's just put aside any attempt to make heads or tails out of the results that relate to trying to interpret the appearance or collapse of the interference pattern. I realize I can't avoid these forever, but can I somehow turn some aspect of this experiment on its head and look at it with a fresh, new perspective that lets me make some sort of headway?

It turns out that the answer is yes. In science, math, and engineering, like just about everything else in life, we can pick and choose among various problems or challenges on our path toward a solution. Some problems, challenges, or approaches are just preferable to others. In math, this concept takes the form of "substitutions" in which we substitute a term with which we are more familiar or comfortable for one that makes our skin crawl. We trade one problem for another and then we see if we can find another productive step to take that might improve our prospect for reaching a fruitful conclusion. Ultimately, mathematicians need to undo the substitution, but only after they've reduced the problem to something that is tractable.

We're going to do something very similar. If we review the countless scores of results for the variants of the double-slit experiments similar to the ones we've discussed so far, we discover something shocking. Every single time we run the test, the results are consistent with the observers that participate in the experiment. There are two ways in which we can phrase this: (1) the experimental results are completely consistent with the observers that are present and participating and for which we can account; and (2) the experimental results are completely consistent with the absence of any unaccountable observers.

Believe me, I understand that this sounds once again like a game of semantics, but it's not. Just as we learned earlier, our perceptions can be deceptive, so it becomes crucially important to define our terms and our concepts. The concept I just introduced at first seems so blatantly obvious that there should have been no reason at all even to mention it. That's not the case, however.

The results of the double-slit experiment give us perfectly clear and unambiguous information, free of any paradox whatsoever, if we analyze the results from a different perspective from which we can acquire clear and unambiguous perceptions. The results are speaking to us in a sense, but we need to adjust the spectrum of our thinking to perceive what information we can.

Instead of thinking about what the experiment is telling us about the world around us (the external universe), let's think about what the experiment is telling us about *us* (the self or internal universe).

From this perspective, the double-slit experiment can be viewed as an observer detector!

Forget for the moment what the experiment is trying to tell us about our universe. What could that matter right now when we can't even define such an important and fundamental term as observer? There is an order to the steps we will take to understand our universe and eliminate the paradoxes, but until I have a way to understand and precisely define what an observer is and the role it plays, how can I even consider building up a thorough and complete understanding of my world?

The concept of the observer is crucial to quantum mechanics and relativity alike. Yet a consistent, common understanding or definition of "observer" is absent from each of these fields. If the concept of the observer is the fundamental element of cosmology, relativity, and quantum mechanics, then how in the world can consistency among those fields and other fields exist and be preserved with any sort of fidelity? Without a common understanding of the concept of the observer, how could it even be possible to avoid the emergence of paradoxes between and among these fields and others? I submit to you that, without a common and consistent definition and understanding of the concept of the observer, it's just not possible to avoid the emergence of paradoxes. I submit that paradoxes arise from inconsistent and disparate perspectives and the absence of harmony.

As I'll soon share with you, there's another term representing yet another concept for which a common and precise understanding is also missing. That second term is sometimes called empty space. Different people, disciplines, and cultures call it by other names, like vacuum, void, or free space. Scientists and mathematicians have developed a host of names for it, but a common understanding has remained elusive. Empty space makes up all but a single part in more than 500 trillion parts or so of ordinary matter, as we've learned. It occurs to me, as I hope it does to the reader as well, that we need to pin down a precise definition of the void. Of course, we'll need some means by which we can achieve that. And as outrageous as it may sound at first, that's the next gift that the double-slit experiment provides us. I'll discuss more about this soon.

Armed with our new insights provided by the double-slit experimental results, let's first pin down some more definitions and enter them into our small but growing glossary:

◊ **Observer** – a living entity capable of perception relevant to a specific observation or set of observations; for example, an observation involving electromagnetic energy requires light perception in the correct portion of the spectrum by the observer.

◊ **Observation** – the synthesis of information by a living entity that connects self-awareness with specific information it perceives about the internal self or the external universe.

◊ **Self** – a living entity capable of conceptualizing a boundary distinguishing that which it considers internal ("itself") from that which it considers external ("the world" around it).

We now have six entries in our glossary; we only have a dozen more terms to add. There are only a few concepts that need to be carefully and deliberately defined from which we can build up an understanding of our universe and avoid the paradoxes.

I think it's important at this point to consider the reason why scientists in the fields of particle physics, cosmology, relativity, astrophysics, and relativity might have avoided pinning down a definition for observer, a concept which is so deeply and profoundly crucial to each of their fields. The short answer seems to be the nature of culture.

A word of advice: Never underestimate the significance of human culture! Culture is an important thing with many positive virtues, to be sure. Culture provides us with identity and a connection with both our past and our future. It connects us to others in our community. It helps us understand who we are and what we are not. Culture helps us feel proud of whom we are—or at least it should. Culture provides us with richness and diversity. Culture makes human society interesting and keeps boredom at bay.

But as we're probably each aware, culture has a dark side, too. People actually look down on other people just because of differences in culture. People kill other people because of nothing more than differences between cultures. Culture provides a way of separating "us" from "them." Culture provides a means by which our biases take greater control of our thoughts and actions than they should, or perhaps more than we desire. Culture provides a way to look at "them" and to make "them" seem horrible.

On the positive side, the culture of the "hard" sciences, like physics, chemistry, and quantum mechanics, is to adhere to the greatest extent possible to scientific methods and "hard," scientifically based, measurable phenomena and repeatable experiments. But we don't have to search for too terribly long or hard before we discover a related, but unfortunate consequence of the objective culture. Don't get me wrong: What I'm about to discuss is not unique to the culture of "hard" science. Not at all. In fact, it's quite the contrary; every coin of human culture has these two sides. I don't mean to pick exclusively on the hard scientists; it's just that it's particularly relevant at this point in our discussion.

The culture of the "hard" scientists is such that it never, ever wants to be confused with the "soft" or "softer" sciences. In fact, I'll have to research this more, but I wouldn't be too terribly surprised if it were the "hard" scientists who even came up with the terms "hard" and "soft" science. The "soft" sciences include disciplines like biology, psychology, anthropology, linguistics, and speech language pathology.

There are a few things which seem to separate hard science from soft science. For one thing, soft science tends to be more humanistic in its nature and practice. It matters how subjects think and feel, and what connections between mind and body may form within their subjects. And, although statistics plays an important role in each of the scientific disciplines, be it hard or soft, complex mathematical concepts, schemes, and models are routinely devised and used in the hard sciences which is less likely the case in the softer sciences. But I think the most important distinction between the hard and soft sciences is that the hard scientists want the soft sciences placed squarely between them and anything that might remotely resemble religion, spirituality, or metaphysics.

This is especially true, I think at least in part, in this new millennium, as rising healthcare costs and a largely apathetic, production-based healthcare system in the United States has motivated individuals to seek alternative and holistic approaches to health care.

The ironic thing is this: Nearly every study performed to help us better understand our world from the perspective of just about any scientific discipline has demonstrated to us again and again the increasing significance of the individual and the mind–body connection. This is true for hard science as well as soft science. After studying the topic as part of my research for this book, I'm left with the distinct impression that hard scientists as a culture fight harder against embracing the role and importance of the individual just as the very importance of the individual becomes increasingly established. I do want to temper this comment with the simple statement that it would be terribly unfair to assign these attributes to all practitioners of hard science.

Since I may appear to be bashing the hard scientists, I do want to remind you that I, too, have been a member of this culture, and heaven knows I've done my share of engaging in hard science snobbery. Along the way, I moved to New Mexico and my perspective broadened as a result.

Let me try to describe a few unique aspects of New Mexico. You may or may not be aware of the peculiar fact that New Mexico has more Ph.D.'s per capita than any other U.S. state. There are laboratories all over the place: Sandia National Laboratories, Los Alamos National Laboratory, Air Force Research Laboratory, and the educational research institutions of New Mexico (University of New Mexico, New Mexico Tech, and New Mexico State University), to name a few. The site of the detonation of the first atom bomb, Trinity Site, is in New Mexico.

Spaceport America is located in New Mexico. This is the place where Virgin Galactic hopes to launch the first space tourism flights. You've got Roswell, and the entire space alien culture that goes along with it, located in New Mexico. Television and movie studios have selected to portray New Mexico as this bizarrely enigmatic place where supernatural and mysterious things happen routinely.

You've got the artist communities of Taos, Madrid, and Jemez Springs. You've got the mysteriously vacant cliff dwelling ruins of the Anasazi in Chaco Canyon and elsewhere in the state. You've got Native American pueblos, art, and culture. Of course, you also have Santa Fe, home to many celebrities, with its rich art and history. The bottom line is this: New Mexico is such a delicious hodgepodge of cultures and beliefs that a physicist can actually openly expose and express a spiritual side and not be made to feel uncomfortable. It may not be a coincidence that the pieces of this book came together only after I spent fifteen years living in this "Land of Enchantment."

One last thing. I mentioned that the culture of hard science includes, among its own rich diversity, a tendency to look down upon softer science and its practitioners. This is not uncommon among the disciplines of hard science. I spent my first career as a semiconductor physicist. I recently had a jarring conversation with a professor of particle physics from the nearby university. I innocently entered into a conversation with this individual about the paradoxes of the universe and some perspectives I've developed to overcome them, which is, of course, the supposed focus of this book. Without any provocation on my part, I noticed that this individual become enraged as I described my view that paradoxes about the universe arise from flawed and/or incomplete scientific understanding. I encountered, much to my chagrin, that he played hard science one-upmanship on me to discourage me from entering into a dialogue that he considered to be the sacrosanct domain of particle physicists. When he failed to discourage me, he attempted to use mistruths and condescension to discourage me. When I surprised this professor by exposing the faulty statements upon which he reluctantly agreed to stand corrected, he was left with no weapons in his arsenal with which to base any further scientific attacks. He just called me a name and walked away at his family's urging.

Einstein once said, "Great spirits have always found violent opposition from mediocrities. The latter cannot understand it when a man does not thoughtlessly submit to hereditary prejudices but honestly and courageously uses his intelligence."

Quite frankly and somewhat embarrassingly for the science and technology community, this really wasn't an isolated event from my perspective. Throughout my career, it has not been unusual in my experience that other scientists, technologists, and engineers have responded negatively and condescendingly about solutions I've proposed when encountering technical challenges. Like any other culture, what is accepted and known is comfortable, the "known devil," while new thoughts, concepts, and approaches represent new and somewhat scary things. The bottom line is that, had I not stuck to my guns and acted on my thoughts, there would still be many problems in my particular field that would have remained unsolved.

In each and every case, the following saying applied exceedingly well:

I've heard it said that truth enjoys only a brief moment of victory between the time it is condemned as paradox and the time it is embraced as trivial.

As humorous as this statement sounds, once you've let it settle in, it actually provides important insights about human behavior, thought, and culture. When a new idea encounters the culture from which it emerges (and I'm fairly confident that it doesn't matter which culture you choose; this probably holds true for any culture), the first reaction of its members is to say that the new idea is nonsense, that it can't possibly be true. The usual argument, as ridiculous as it sounds, is that if the new concept were true then it would already have been accepted as truth by the culture. If you spend just a moment thinking about this, then that notion violates our concept of causality and it could never, ever be true. After all, how could a culture embrace a concept before the introduction of that concept to the culture?

Anyway, the second half of the quote is the more humorous and telling aspect of it. There's a very brief moment during which a very few colleagues will say something like, "good job," to provide some meager feedback on the contribution that the idea now represents. After this very short time, others in the field simply accept and embrace the new concept like it's been there for all eternity, as obvious as the nose on your face. You know that you've made a meaningful contribution when you hear colleagues say things like, "Of course, that was the solution. In fact, that was an important element of the research I was doing nearly ten years ago."

That sort of experience seemed frustrating to me at first, although the humor was never lost on me. But, after a while, I embraced this sort of experience as the validation that my thoughts, inventions, and contributions were on target and that I made important contributions to my field. I'd even go further and say this is the outcome you better hope for if you want to make a meaningful contribution within a field or culture. This outcome, more than anything else, says, "good job." Viewed from the right perspective with the right attitude, everything about these experiences should be considered good—very good.

I have a few more words about the culture of science and technology before we move forward in our quest to eliminate paradoxes. There are just a few ironies in the field of science and technology that I'd like to share with you to help broaden perspective and to expose some of the embedded biases.

Again, going about a usual day in New Mexico usually exposes you to a staggering spectrum of human beliefs and cultures. Within the demographics you'll find in New Mexico, the scientists and the spiritualists constitute two very prominent groups. As each of us is likely familiar, the concept of crystals is often a central theme of spirituality. Spiritualists consider crystals as having special powers and properties, and crystals are used extensively in the procedures and rituals of spiritual practitioners.

The science and technology culture, as we discussed earlier, possesses certain rigid attitudes—once again, not an uncommon feature of human cultures, good or bad. Given the proximity and intermingling of these two cultures in New Mexico, it's not terribly uncommon to hear a scientist or engineer mocking spiritualists with respect to their views on crystals. I know this because I've participated.

Then I had a startling revelation one day that there's something deeply ironic about scientists and engineers making fun of the importance of crystals to spiritualists. My entire career in science and technology depended entirely on the unique aspects of crystals! I worked with radiation effects in semiconductors. Every single activity during my career involved in one form or fashion the special properties of crystals. Crystals were the very substrate upon which I was developing space computing solutions. I used crystals for measuring the amount of radiation to which I exposed my test circuits. Without crystals, my field wouldn't even exist!

Crystals of semiconductor materials represent the foundation of today's computers and are unequivocally the basis for the Information Age. When I think about it, there are few fields in hard science that don't rely in some important way or another on crystals.

So, in retrospect, there are at least two communities that place profound importance on crystals and their special properties, and these are the science and the spiritual communities. Needless to say, I no longer engage in mocking spiritualists ever since I pulled the rug out from under my own feet.

But now let's talk about a more significant bias within the science culture. There's a common goal to which the hardest of the hard sciences aspire. The goal has a name: the Theory of Everything (TOE). This goal has dominated the work of hard science for the last several decades. Einstein unsuccessfully devoted his final years to this goal. Among others, String Theory, and its newer variant, M-Theory, emerged as a new approach toward the development of the Theory of Everything. Some of science's most costly facilities, laboratories, and experiments, like the Large Hadron Collider at the European Organization for Nuclear Research, are ultimately devoted to playing an important role in the development of the Theory of Everything.

As its name implies, the Theory of Everything is pretty all-encompassing. I've mentioned at times throughout the book that science has identified four fundamental forces at play in our universe. And, although there's some speculation of a new force called dark energy, we're going to ignore dark energy for now and return to it at a more appropriate point in the book, when we have better context for it. Scientists have succeeded in developing an understanding of the relationships among some of these fundamental forces but not among them all. The ultimate understanding of the fundamental relationship among all four of the forces, which has thus far eluded scientists is the Theory of Everything.

Let's review what those forces are because we've discussed most of them already in one form or fashion. There's the strong nuclear force (also known by its shorter form, the "strong force"), which keeps protons together in the nucleus of an atom and operates over very short distances. Then there's the weak nuclear force ("weak force"), which keeps neutrons bound in the nucleus of atoms and also operates over very short distances. There's the electromagnetic force with which we are very familiar, of which things like radio waves, microwaves, and visible light are comprised. The electromagnetic force operates over very great distances, which should be obvious when we glance upward in the nighttime sky to look at the stars. Their light reaches us from vast distances. As light emanates from a point in space, the light expands outward in all directions and therefore diminishes by the square of the distance between a light source and any particular point in space. (Collimated beams of light like those produced by laser pointers don't follow the same relationship since the light they produce doesn't spread outward in a sphere.) Finally, there's gravity, also known as the gravitic force. Gravity, like the electromagnetic force, diminishes by the square of the distance between two objects since it, too, is projected outward in a sphere.

One more note about gravity: It is outrageously weak compared to the other forces. In this and other ways, gravity is sort of the "odd man out." Here's a quick illustration of just how weak gravity is. Find a magnet and use it to pick up a paper clip. Chances are you found some little magnet in some utility drawer in your home or you grabbed a magnet off the fridge that was holding a note, or a photograph, or some silly piece of art your child produced in grade school. Now look at the situation in front of you. Your hand is holding a tiny magnet that's lifting a paper clip against the combined gravity of the entire planet! There's a huge difference between the magnitude of gravity and the other forces and it is hopefully abundantly obvious to you now.

I've dropped some hints already that gravity is the outlier. The relationship among the other three forces is largely understood. Scientists would say that the strong force, weak force, and electromagnetic force have been "unified." The theory which unifies these forces is referred to as the Grand Unified Theory.

There's also a related term with which you might have come into contact at some point, called Unified Field Theory. Unified Field Theory is a term that usually refers to the attempt by Einstein to unify relativity theory with quantum mechanics. The result of such an endeavor is essentially equivalent to the Theory of Everything, since quantum mechanics focuses on strong, weak, and electro-magnetic forces, while relativity focuses on gravity and electromagnetism.

Hard scientists in the fields actively engaged in developing the Theory of Everything take for granted that all four of the forces will ultimately unify into a single theory and with it a single set of fundamental mathematical descriptions that describe the relationships among all four of the forces. In essence, the result would be a single mathematical description that underlies all physical phenomena in our universe and the reality contained within it.

Scientists have worked arduously since the middle 1900s to develop the Theory of Everything and have thus far been fruitless. Not only is there no Theory of Everything at the time this book was published, but there is no credible prospect of developing this theory in the foreseeable future. What's worse is the absence of a single shred of empirical evidence that the four fundamental forces can unify into a single theory. Enormous amounts of energy, effort, brainpower, and dollars are being poured into the effort.

Let's take a step back once again and view the situation I just described. There's a belief within the science culture that the four forces can be unified. Again, there's a complete absence of empirical proof upon which to base this belief.

Now, add into the mix the fact that today's science introduces a number of paradoxes whose existence points to a flawed and/or incomplete set of scientific interpretations.

Finally, add into the mix that today's scientific doctrine is strenuously defended by its practitioners while they make disparaging remarks about the science of other practitioners, and an ugly picture emerges. At the very heart of today's hard scientific endeavors to understand the universe are beliefs and a culture indistinguishable from religion.

By these measures, perhaps the only thing upon which hard science holds a monopoly is at its very heart a religious belief.

There was an Indian philosopher by the name of Nagarjuna. He very strenuously taught that the greatest risk arising from any particular, compartmentalized school of thought was that its doctrine would ultimately transform into dogma that its members would mindlessly defend. The risk of adhering to any single, compartmentalized school of thought is the possibility of missing opportunities to grow and thereby exclude other important concepts and perspectives that might prove useful.

I, too, believe that the four fundamental forces will ultimately be unified into a single coherent description of the universe. But, I further believe that the path to achieve this must include a good understanding of the role of the observer and an understanding of the individual that includes important and fundamental roles for the softer sciences. I believe that understanding the universe can only be achieved when science comes to grips with the significance represented by the individual. Perspective may turn out to be everything, and the only entity in the universe possessing perspective is the individual, and the only place where that individual's perspective exists is within its mind. The mind of the individual observer is truly the bridge between us (our internal universe) and our world (the external universe).

Now, let's get going! We've got problems to solve and we're going to need open minds to proceed in the most fruitful way.

Chapter 5

Return of the Observer Detector

So, we're going to return to the double-slit experiment with our new, fresh perspective that the first message to glean from the double-slit experiment is not about the external universe; rather, the first insight we gain from the double-slit experiment is about the nature of ourselves, about our identity as an observer. After some conscious interpretation, we see that the experiment, in fact, informs us that we're observers.

After thirty-four years of pondering what the darn double-slit experiment was telling me about the universe, this realization came as quite a shocking revelation. My first reaction was to question this insight, to return to the voluminous amount of information about the double-slit experiment and see if I would once again derive the same thought. As I did so, the concept began to really take hold, especially since it provided me with a fresh perspective by which to move forward. So I established this thought as my first hypothesis and decided to travel down the road a bit further to see what developed.

Hypothesis #1: A double-slit experimental apparatus is first and foremost an effective observer detector.

As good scientists, we will now set out to disprove our hypothesis and, if we come to some impenetrable obstacle, we'll have to abandon the notion and consider the hypothesis to be false. (I have the benefit of knowing how the book ends, so let me just share with you now that the hypothesis has withstood my scrutiny; so I've accepted the hypothesis as true.)

For the double-slit experimental apparatus to serve as an observer detector, some things must always be true:

◊ The experimental apparatus must respond in some measurable way to the presence of an observer and provide a positive response.

◊ The experimental apparatus must respond in a repeatable fashion such that it responds identically to each observer or group of observers with similar or identical observational capabilities.

◊ The experimental apparatus must never respond in the absence of an observer as it would in the presence of an observer.

◊ The experimental apparatus must respond in a repeatable fashion such that it responds in an identical way all the time to the absence of observers.

There are some responsibilities placed on us as the experimenters:

◊ We must share with others the precise nature of our experimental setups so other researchers can repeat our experiments.

◊ We must faithfully record our observations.

◊ We must be able to account for each observer or the absence of observers. In other words, we must be able to clearly specify when observers are introduced or removed from the experiment.

◊ We'll need to analyze our experimental results and report them to others.

We can benefit from the extraordinary number of researchers who have already performed the experiments for us by taking advantage of the abundant information provided in the scientific literature. Allow me to summarize the results I've researched in the open literature:

◊ Without exception, each time an accountable observer[3] attempted to identify or otherwise determine (measure) through which slit a particular photon traveled, the interference pattern collapsed.

◊ Without exception, each time an observer did not attempt to perform this measurement, the interference pattern was preserved.

◊ The interference pattern never collapsed in the absence of accountable observers performing a measurement.

◊ The interference pattern never collapsed when an accountable observer was present, but not actually performing, a measurement.

◊ The interference pattern was always present when measurements were not being taken or when no *unaccountable* observers were present.

The results are unambiguous. The experiment provides consistent, repeatable results. We can now accept our hypothesis as true and proceed onward.

As usual, though, let's spend a little more time to consider these results. Despite the fact that these were the results we were expecting, there are a couple of noteworthy comments to add.

Even though the purpose of our experiment was to examine the nature of the observer and the individual, I really never felt like we crossed some line into the metaphysical or supernatural. In fact, I was quite relieved to discover that the experiment was consistent with my disbelief in such things. (I really don't want to harp too much on the next assertion I will make based on the experimental results. But it's my book and I will make this statement if I want to!) Since the results were completely consistent with accountable observers, we can make a scientifically based assertion that either no phantoms wandered across any of the experiments, that phantoms are incapable of achieving observer status, and that there is ample scientific evidence based on the double-slit experiment to completely discard certain unfounded, supernatural beliefs about the deceased.

This will be the one and only time in this book I ever make any judgment about religious or supernatural beliefs, and I just know I'm opening myself up to commentary and criticism. Nonetheless, my comments are based upon scientific evidence and that's that. In case anyone was wondering, I do subscribe to religion, but I desperately try to keep it separate from science, even though I don't always succeed.

There's one other thought I want to share with the reader. Beginning with our study of the double-slit experimental phenomena, I get the distinct impression that the observer alters the universe in unimaginable ways as he constructs a bridge between phenomena in the external universe and his/her subsequent perception, interpretation, and understanding of it. I feel that it becomes more and more clear to me that the relationship between the internal universe and the external universe is as indistinct as the boundary between my fingertip and any object I attempt to touch.

Now, let's take a moment to review what another scientist has shared about the double-slit experiment and his insights that relate to it.

It's time to bring up Richard Feynman again. Dr. Feynman believed that the double-slit experiment provided a unique window from which to understand our universe and the quantum mechanical theories that attempt to describe it fully and accurately. In fact, Dr. Feynman thought the double-slit experiment was so important that he believed, ultimately, a complete understanding of the universe could be derived from it along with a complete and thorough quantum mechanical model. He made this assertion decades ago, before his death in 1988.

I often read and reread Dr. Feynman's bold statement and wondered if it could possibly be true. It's among the main reasons, in fact, why I spent so much of my time and energy focused on understanding and gleaning what I could from the variants of the double-slit experiment. And, trust me, there are variants of the double-slit experiment that I'll introduce in subsequent chapters that will positively blow you away if this is the first time you're encountering them. I certainly encourage you to take the initiative and perform additional literature searches and read other books on the topic in order to assist you in comprehending this amazing universe of ours. The book's Web site offers a variety of links that you may find useful.

Dr. Feynman was convinced that unique insights into the nature of the universe would come from nothing more than analyzing and understanding the results of double-slit experiments. Needless to say, he was intrigued to discover what he could from these experiments. As a positively brilliant mathematician, he explored mathematical methods to comprehend the results. Among the results for which he is famous is his Sum-Over-Paths Model.

Chapter 6
Feynman's Sum-Over-Paths Model

Dr. Feynman recognized that quantum mechanics could not tell us the precise path that a particular photon takes on its way from the light source to the screen. Quantum mechanics uses wave descriptions of particles to determine the probability of finding some particle within some region of space. Recall that the Heisenberg Uncertainty Principle tells us that the more precisely we know about certain characteristics of a particle, such as the position of a photon, the less precisely we will know about other characteristics, like where the photon is headed to next. It is through wave functions that quantum mechanics describes particles of matter and the mathematical constructs that we associate with the particles that serve as the carriers of force, like the photon.

The wave function of the photon can give a scientist a good understanding of the probability for finding the particular photon in a particular location, but it can't tell the scientist where to actually find the photon. Nor can it inform the scientist of the path that the photon chose to take through the universe.

Well, Dr. Feynman saw the interference pattern and must have thought (rightly, in my humble opinion) that the interference pattern must indicate the interaction of multiple wave functions. Dr. Feynman set out to see what the interference pattern would look like mathematically if the photon in the single-photon variant of the double-slit experiment were to take every conceivable path through the universe simultaneously. (Recall that in the second variant of the experiment, we only allow one photon to be en route to the projection screen at any one time, and only well after the photon en route has been detected on the screen do we let the next photon rip forward into the apparatus.) The result is his Model of Sum Over Paths.

The interference patterns generated by his summed paths mathematical model are indistinguishable from the summed results of the single-photon variant of the double-slit experiment! Wow!

It completely defies any sense of understanding that we might base on our everyday experiences, but that's not uncommon when it comes to quantum mechanics and the quantum universe. In fact, the reader should be getting somewhat accustomed to being stunned by this point in the book.

If Dr. Feynman's results represent an accurate understanding of the double-slit experiment, then the universe must be inconceivably different than what we ordinarily imagine it to be. After all, it seems inconceivable that a photon could take every path through the universe and arrive in a timely fashion to be included in our observations. As you'll see in discussions of other variants of the double-slit experiment, our concepts of time and space will be called into question. Keep this discussion in mind. We'll be returning to it later.

Let's return once again to where we left off in our own analysis of the double-slit experiment and finish what we started, armed with a better understanding of the concept of the observer.

Now that we've established ourselves as observers, is there more that the double-slit experiment is trying to tell us? Surely, I wouldn't have asked this question unless the answer had been yes!

If we observe an interference pattern in the single-photon variant of the experiment, what else are the results indicating to us? If an interference pattern results from the interaction of the wave functions of multiple photons, as we observed in the variant of the experiment we performed earlier, then our internal logic might suggest to us that there is at least some possibility that other photons and wave functions are participating in our experiment without our knowledge or consent.

In essence, we just encountered a fork in our road. On one side of the fork lies the possibility that, indeed, we had only a single photon and wave function, in which case our result appears to be not interpretable and leads us to the paralysis that's persisted in scientific understanding since its discovery. On the other side of the fork lies the possibility that other "photons" were present despite the scientifically based assertion to the contrary. Hmmm. This does indeed seem to be a problem.

Only in the world of quantum mechanics can we simultaneously consider the possibility that each fork can be both true and false, and that there is actually a meaningful path forward. This is going to involve a great deal of explanation on my part, so here it comes.

Recall that a basic premise of quantum mechanics embraced within the Copenhagen Interpretation is that nothing is considered to objectively exist unless its existence is revealed through observation and measurement. Yet another basic premise of quantum mechanics is that only upon observation does an infinite set of possibilities collapse down to the single value that the observer measures.

Each of these concepts is absolutely fundamental to the field of quantum mechanics and each is embraced as truth by the quantum mechanics and particle physics communities. I see a bit of a disparity here that I can exploit to find a middle ground. I'm not choosing either of the paths presented to me by our dilemma; I'm going to cheat and take both paths and neither path simultaneously.

Let me explain how I intend to do this. There is a loophole that exists in the scientific method. Sure, by carefully combining a good hypothesis with the careful and deliberate planning and execution of a good scientific experiment, a scientist can support the truthfulness of one possible understanding of a physical phenomenon. But, the scientific method is not capable of proving the non-existence of a phenomenon or that an alternate understanding of the phenomenon does not exist. Science can only wait for a disparity to form between its current theories and understandings before setting out to develop new science or new physics.

For example, the decay characteristics of muon particles in Earth's atmosphere and the characteristics of Mercury's rotation could not be understood in terms of the science of the day, at the turn of the twentieth century. Einstein's relativity provided new physics to finally explain these paradoxes while maintaining consistency with prior scientific understanding.

So, science can confidently tell us that it detects no other photons involved in our experiment. Science, however, simply cannot prove the absence of other photons. Is it possible, then, for other photons to indeed be present without them informing us of their presence and without stepping on the toes of science?

The answer, I believe, is yes, and my explanation will shed some light on interpreting the results of any variant of the double-slit experiment.

Let's start by making a distinction once again between the external universe and the internal universe. We've spent quite a bit of time understanding the basics of how we perceive the universe.

We perceive the universe through perceptions. We each have ample experience in understanding how deceptive our perceptions can be. For instance, we discussed how our fingertips don't actually make physical contact with objects and yet the sensation of touch so strongly suggests to us otherwise. We discovered that the volumetric difference between the object we "touch" and empty space is only one part in about 500 trillion. Our perceptions certainly send us deceiving messages, yet the messages sent to us through our senses served our purposes quite well over time.

Our senses and perceptions don't seem adequate or sufficiently dependable, however, to be of much use to us when it comes to our quest to understand a universe free of paradoxes. On one hand, of course, our senses and perceptions are our best friends. They tell us that there is in fact a universe to ponder. They tell us where to find nourishment and when it's time to seek it. They enable us to seek mates and to maintain the continuity of humanity. They enable us to function in our daily lives. We derive a great deal of benefit and gratification from our senses and perceptions.

But, on the other hand, our senses and our perceptions can be devious, deceptive imps. For one thing, our senses operate over frustratingly small portions of the possible range of values or spectra actually available in nature. The best functioning human eyes can detect only a vanishingly small fraction of the electromagnetic spectrum and, without the additional assistance of optical tools and other technology, human eyes can only function well over frustratingly short distances. The best human hearing detects acoustic vibrations over a frustratingly small portion of the spectrum of "sounds" which are produced in nature. Our senses of taste and smell operate over such a small range of possible chemical signatures that many of nature's fragrances and chemical signatures remain undetectable to us, relegating them to tasteless or odorless phantom-like substances, with which we will never cultivate a relationship. Our sense of touch is the most limited in its range of operation; our range of touch extends only as far as our limbs. Our sense of temperature is essentially limited to two values—heat in or heat out—although the added dimension of the rate of heat flow provides us with additional insights designed to help us avoid dangerous situations. Our sense of touch is indeed pleasant, but it operates over an exceedingly small range of pressures available in nature before the sense of touch produces the sensation of pain.

We've learned that our senses and perceptions provide us with a very definite and very deceptive concept of boundary. We possess a dramatically misguided notion of a firm, fixed, sharp boundary between us and the world around us. Similarly, we possess an equally misguided notion of a firm, fixed, sharp boundary between our minds and our bodies and between our minds and our external world.

Yet, the "double-slit experiment as observer detector" begins to provide us with unique insights about the nature of the observer and, with it, the nature of the universe. We've unambiguously and consistently observed that the simple act of performing a measurement changes the universe in a clear and demonstrable way. The simple act of performing a measurement causes the interference pattern to collapse. The collapse of the interference pattern indicates to us that we are an observer. It provides us with insight into the nature of the quantum world. As a consequence, the collapse of the interference pattern has caused changes in our understanding, in the way we think, and in the relationship between us and the external world.

Our interaction with the quantum world altered the external universe and changes to the external universe altered key aspects of our internal universe.[4] These cascading actions and the resulting perceptions altered my perception of the universe and the concept of my role within it. And the changes that occurred within the external universe and the internal universe are bridged by my mind—the same mind in which I store my interpretations and experiences. Observers alter the state of the universe, and the universe alters the state of observers. Pretty freaky, huh?

At this point we would be perfectly fine if our experiences with the observer detector were the only thoughts stored in our minds; unfortunately, that's not the case at all. Our minds were not intended originally to store and process observations upon which to understand a universe without paradox. No, they were intended to help us survive. So, we have to face the unfortunate fact that we have a lot of other stuff stored in there, a lot of which is just simply irrelevant to understanding the universe, and a whole bunch of stuff which, worse, obstructs our ability to understand the world. Things like false perceptions, biases, wrong or misplaced concepts of the self, anger, pride—all the stuff our parents succeeded in putting into our heads—fears, and so on, just litter the landscape of the mind.

So, we humans try to sort it out. We try really, really hard to devise a set of mechanisms and rules that help us sort out the "real" information from the phony or useless stuff. And so, philosophy was born nearly three millennia ago. Philosophy based on purely abstract reasoning emerged with the pre-Socratic philosophers about 2,500 years ago and the seeds of the modern scientific method were sown. The first saplings of "modern" science began to sprout as early as about two millennia ago.

Yet, throughout this time and ever since, philosophers and scientists remained quite familiar with the simple fact that perceptions are already biased by the time we perceive, and subsequently store, them for later use. The very act of perception is a grotesquely skewed and biased process. Our minds don't work in a way that permit us to record raw data and to retrieve them later for interpretation, as our scientific instruments are designed to do. The very act of perception introduces fallacy that is implausible to remove later. By the time we have perceived, we have already prejudiced the information we store in our minds.

And so, we don't see the world accurately whether we are scientists who fancy to faithfully emulate Spock or whether we are poets. The Jewish Talmud, written between 200 and 600 CE, captured this concept well. The Talmud says, "We don't see things as they are; we see them as we are."

We devised an imperfect scientific method, to be sure, but it's way better than not having it. The scientific method has permitted humanity to make increasingly great strides, sometimes to the betterment of society and sometimes not. Regardless, scientific progress marches on.

Aristotle established the original school of physics, and despite its falla-cies and inaccuracies, Aristotelian physics stood as the final word in its field for nearly two millennia. It was only with the emergence of Newton and his school of physics that Aristotelian physics met its demise. Newtonian physics became the final word for more than two centuries. Although the number of paradoxes unexplained by Newtonian physics was small, it increased over time with increas-ing observation and precision, and these paradoxes took on added significance. Finally, at the start of the twentieth century, Einstein provided the world with its latest final word on physics, the scientific world's third school of physics—Einsteinian physics—and the combined theories of relativity. Within a meager couple of decades, a new school of physics emerged known as quantum mechanics, and the sparks really began flying.

Relativity makes astounding claims about the nature of the universe that subsequent experimentation bears out. Relativity is right on target. Quantum mechanics also makes astounding claims about the nature of the universe, and again, subsequent experimentation demonstrates that quantum mechanics is right on the money.

Isn't that fantastic? Shouldn't scientists worldwide prance in exuberance about these impressive accomplishments? Well, there's a problem. Relativity works fantastically well over astronomical distances, and quantum mechanics works fantastically well over puny distances. But they provide outrageously inconsistent predictions when you try to mix and match them over the full spectrum of the universe. The scientific predictions of the theories become outrageously disjointed with scientific observations of the "real" world.

Relativity and quantum mechanics are today's final word on physics, and the number of paradoxes they produce is staggering. The rate at which new para-doxes are emerging is equally staggering. As I mentioned earlier, the summed mass-energy of the universe has increased twenty-five-fold over the last decade alone as a result of new and improved scientific measurements and technology.

Here's the point. When it comes to paradoxes, you really only have two choices: (1) discard or ignore the offending data, or (2) improve the understanding. The scientists who subscribed to Aristotle's physics focused on the former strat-egy and that really didn't work out too well for them or the world, in retrospect. The reason was that Aristotelian physics transformed over time into religious dogma, both figuratively and literally. People and institutions forced scientists to denounce new concepts and scientists were put to death for nothing more than following the scientific method, taking measurements, and trying their best to make heads or tails out of their experiments and interpretations. It was tragic and unnecessary, and it had a profound impact on the pace of scientific advancement.

Chapter 7

Paradoxes and Doctrine:

A Lovely Couple

Now, here we are today in 2011, and it's been decades since a true scientific breakthrough has altered our understanding of the universe. Might the reason for this be that relativity and quantum mechanics have written the final word on the physics of our world? No, we can be fairly certain that is not the case. Might the reason involve a paradox shortage? No, we can be fairly certain that is not the case, since paradox collectors are blessed with ample collections at this particular time. Might the reason be that the low-hanging fruits have all been harvested and now we need really costly, special ladders to continue harvesting the tougher-to-reach science?

I'm reasonably confident that this isn't the case, either, even though science and its public presence revolve more and more around really big, costly projects, such as the Large Hadron Collider, the Laser Interferometer Gravitational-Wave Observatory, and a new crop of space-based science experiments. I think it's far more likely that outrageously expensive science experiments are really more of a symptom than an underlying cause of the problem. The Laser Interferometer Gravitational-Wave Observatory, or LIGO, is a physics experiment which is attempting to detect gravitational waves through laser-based electromagnetic sensors. LIGO is a joint project involving scientists from MIT, Caltech, and many other colleges and universities. The project is sponsored by the National Science Foundation (NSF). At $365 million (in 2002 USD), LIGO is the largest and most ambitious project ever funded by the NSF.[5]

I think the most likely reason is that science has reached the truly unfortunate status of dogma. I believe that our flawed and incomplete science has once again achieved a religious-like state of doctrine, and this time scientists did it for and by themselves without the force applied by religious institutions (which, of course, is the old-fashioned way of constructing and establishing dogma).

Don't get me wrong: I imagine the science community would love nothing more than to achieve some truly grand new success, a breakthrough that heralds the age of the next school of physics, bringing us yet closer to the true, final understanding of the universe captured in an honest-to-goodness Theory of Everything.

I think many factors have contributed to this situation, and none of them actually have anything to do with science. Rather, they have to do with human culture, once again. There's the increasing specialization and compartmentalization of science that's been occurring over the same time as our increasing understanding that such compartmentalization is synthetic, unfounded, and unproductive. We've also witnessed the emergence of scientific elitism, the likes of which I've only touched on earlier. This gives rise to a cultural attitude among various scientific communities that some science is "harder" than other sciences, an attitude that can be summed up as "My science can beat up your science." This attitude appears to me to be largely based on the extent and complexity of the mathematics that a particular scientific discipline, subdiscipline, or subsubdiscipline uses. The greater the extent and complexity of the math, then the harder the science is perceived to be by its practitioners and members.

Perhaps, also, the duel between two competing physics camps plays some sort of role. Over the course of a few decades, relativity and quantum mechanics have each had ample opportunity to dig in their respective heels and project undignified portrayals of each other.

Now the situation is that a plethora of variations on a theme have emerged and, much to both my chagrin and disappointment (and I'm sure to yours as well), these variations have blazed brightly for a short while, only to fizzle out again and again. Theories like super-symmetric string theory (otherwise known by its shorter moniker, "String Theory"), and its greater dimensional cousin, M-Theory, have shown some promise in providing a basis for the relationship between gravity and the other forces. Yet it seems more likely to me that these efforts are not leading us in a fruitful direction. My reason for this is simple, along the lines of what my statistics instructor said: Math can be used as a tool by which you can achieve almost any answer you want. By this I don't mean that math is useless or a waste of effort. Quite the contrary; I think math is a crucial endeavor and perhaps the most important tool in the tool chest of science. But math, like any other human construct, can be used in better or worse ways. To disconnect the physical world from the math and pursue interesting mathematical concepts, models, and relationships can become a misguided endeavor, nothing more than a contemporary version of numerology or alchemy.

Einstein said, "As far as the laws of mathematics refer to reality, they are not certain; as far as they are certain, they do not refer to reality."[6]

After all, the tools provided by mathematicians could not have been developed or used previously for intents and purposes which had not yet been conceived. I feel, then, that the circle of related paradoxes that emerge from today's science is at least in part due to the fact that existing mathematical models were employed to ensure internal self-consistency, thus completing the paradoxical circle.

I realize that the last paragraph holds the prospect of me receiving oodles of angry mail from mathematicians. I want to take this opportunity to state that I hold mathematicians in the highest possible regard. Some of my best friends are mathematicians, I will have you know. I'll even go further. Contemporary "hard-science" scientists would be nothing without the tools provided to them by mathematicians, and mathematicians don't get the recognition (or salary) they deserve. Tool makers just don't get a lot of credit. For instance, I have no idea who might have produced the wood plane that Antonio Stradivari[7] used to build his extraordinary musical instruments. History just didn't bother to record it. Too bad.

Okay, so variations on a theme haven't brought us too many terribly new things over the last few decades. Surely, we must have gotten something new. Actually, we have, and some aspects of these new or reinvented concepts hold out some prospect of being useful for creating new perspectives on the way we view our world. This group of new things includes ideas as diverse as the Anthropic Principle, biocentrism, the multiverse, and a universe of higher dimensionality. Again, I don't mean to sound too critical about these concepts. Each one offers a fresh, new perspective with the potential to initiate a new way of thinking. It can only be viewed as a positive thing when new thoughts and concepts emerge and are genuinely shared among the community of humankind. Some of these concepts even emerge from hard scientists and appear as spiritual as just about anything I've ever encountered in New Mexico. And, once again, this is great because we can be sure of just one thing: Our current understanding of the universe (both internal and external universes) is flawed and incomplete. It may be better than what we had before, but it's wrong, nonetheless.

I would also like to add that reading about Dr. Lanza's biocentrism opened my mind in immeasurable ways. In fact, I developed my concept of the double-slit experiment as an observer detector on the very same day I read Dr. Lanza's book. I wrote this entire book in a three-week period—two months after reading Dr. Lanza's book.

Alrighty then: Are we left with a reasonable and meaningful path with the prospect of leading us to some new enlightenment? Yes, absolutely. And this path may be a whole lot simpler than we might have expected.

The way I see it, there are two parts of science, as I think and hope most of you will agree. The first part consists of developing hypotheses, performing observations, taking measurements, and recording raw data. The second part consists of analyzing and interpreting the data, and drawing conclusions. The former part largely involves developing and implementing technology to objectively collect raw, unadulterated data about our universe. The latter part largely involves scientists harvesting nuggets to support their hypotheses from the large, growing, flawed, and incomplete scientific database comprised of the possibly flawed and incomplete conclusions of their predecessors.

So, in essence, I'm confronted with a choice. On one hand, I can assume rightfully or wrongfully that the data portion is unquestionably untainted and accurate, even if incomplete. (I refuse to assume that we've observed all there is to the universe!) On the other hand, I can go with scientific interpretations provided by the current state of human science and assume rightfully or wrongfully that modern science has faithfully and accurately interpreted the data brought to it, free of bias and flaws, even if incomplete. On the third hand, I could accept both the data and the interpretation as accurate, unbiased, and unflawed, even if incomplete.

Hmmm. It really doesn't seem like much of a choice to me. I'm going with option one: a good, even if incomplete, collection of unbiased, accurately measured, and faithfully recorded data. It's just not possible for me to choose either of the other two options, given the history of science and human bias.

For the purpose of this exercise, then, I choose to believe that our paradox collection emerges from faulty and incomplete scientific interpretation, and that paradox is not an innate state of our universe. Although this sounds more cynical than I really mean, given the choice between listening to the universe or listening to human interpretations, I'll choose listening to the universe. Believe you me, the irony of this statement is not lost on me, given the fact that I've written a book to try and convince you of my interpretations regarding the universe. Keep in mind, though, that I'm not asking you to believe one particular set of conclusions; instead, I'm asking you to examine and evaluate for yourselves the methods I use in reaching my conclusions.

So this approach will become the basis of that which I present through the remainder of the book. There are certain ground rules, or boundary conditions, that arise as a consequence of the assumption I've chosen. Anything I say must be completely consistent with scientific data, but there is no such requirement placed on me regarding being consistent with scientific interpretation.

As I went through this exercise on my own, I tried to carefully select my battles. My goal was to identify only a few faulty scientific conclusions and/or interpretations that would enable today's scientific paradoxes to disappear. I used a form of abstract logical reasoning first developed by pre-Socratic philosophers called *reductio ad absurdum*,[8] which means that you try to take a hypothesis to some completely absurd conclusion. If this happens, then the next step involves revising your assumptions and your hypothesis and then taking the revised hypothesis toward a new, absurd conclusion. Greek philosophers successfully used this technique iteratively to produce some of humanity's most stunning insights about the nature of the universe. But I possess a huge advantage over the ancient Greeks, for I have well over two millennia of scientific hindsight upon which to draw.

As I performed this process over the last thirty-four years, my results startled me. I discovered only a handful of scientific interpretations that yielded the paradoxes we observe. A framework began to emerge from which the paradoxes simply don't appear.

I've alluded to the hubris of modern scientific culture, so I want to make one disclaimer before proceeding. I am not making the assertion that this handful of scientific interpretations is either correct or faulty. As should be the case in science, I leave this judgment to my peers. They may choose to replicate my logic and determine for themselves whether the insights I propose are valid or rubbish. Nonetheless, I am making an assertion that the thought process I used and the framework I developed are useful in evaluating our perceptions and the scientific interpretations relating to the nature of our universe. I may even be correct in asserting that a handful of key scientific interpretations and conclusions have produced the paradoxes we observe, but the list of faulty and/or incomplete scientific interpretations and conclusions I've drawn may itself be flawed and/or incomplete.

I believe, as William of Ockham did, that the nonessential elements of my arguments must be pared away but that the paring must stop before we remove the essential portions of the argument. I'll talk more about Ockham later, as well, and how his "razor" became an important tool for scientists and philosophers.

Now let's proceed to my second hypothesis and discuss its basis and ramifications.

Hypothesis #2: It is possible for an unobservable phenomena or object to exist in my vicinity.

Whoa! Isn't that a complete violation of a fundamental tenet of quantum mechanics?

Now wait just a minute! We all know that ordinary matter can be and is routinely observed using electromagnetic force. Sometimes we refer to electromagnetism as light—especially when our eyes can perceive it—so, we refer to any matter we observe through electromagnetism as luminous matter. If there were matter surrounding me right now that I couldn't detect—in other words, "non-luminous" matter—you would expect the same sort of nonsensical phenomena to occur elsewhere in the universe. And, if this special matter is non-luminous, then we should be able to detect it by its gravitational influence, right? Huh! What do you have to say about that, Mr. Smarty-pants? Oh, wait a moment: Isn't there a whole bunch of non-luminous stuff out there that we only recently detected that we call "dark matter?" Oh yeah, that. We'll return to this; I give you my word.

My hypothesis completely violates a basic tenet of the Copenhagen Interpretation of Quantum Mechanics, but I'm going to stick with it and I'm going to convince you that my hypothesis is valid.

Recall that the Copenhagen Interpretation asserts there can be no objects, phenomena, or objective realities other than those which can be revealed through observation and measurement. At first glance, this assertion would seem to imply that the absence of a measurement regarding more photons entering our experimental apparatus is equivalent to a measurement confirming the absence of those photons. This isn't the case.

Science does not work symmetrically with regard to proving the presence or absence of phenomena. In other words, science is well-equipped and highly capable of proving the existence of some object or some phenomena. Science can effectively prove, for instance, the existence of four fundamental forces, but it is ill-equipped to deal with the flip-side of the argument. Science can't effectively prove that some currently unknown fifth or sixth force absolutely, positively does not exist. Without some insight into the nature of these unknown force prospects, I can't pull out science and state they don't exist. The best science can say is that it used its capabilities to the best of its abilities and the results do not indicate the existence of these dreamed-up additional forces.

So, let's take another example. Put 6 billion people in a room and the concept of God is bound to arise. What's more, it becomes apparent that everyone doesn't agree on one particular concept or definition of God.

I think that people in general believe that scientists don't believe in God; My experience tells me otherwise. I think there are some scientists who believe in God; I think there are some scientists who don't believe in God; I think there are some scientists who just haven't decided; and I'm sure that there are other scientists who could be clumped into another bucket we just call "other." In fact, scientists are just like people on the whole.

So, within this distribution, there are very likely scientists who would like to prove that God doesn't exist. These scientists might want to rain on everyone else's parades with respect to God really, really badly. The problem is that one of the limitations of science is that science simply cannot prove the non-existence of objects and phenomena over the full spectrum of possibilities.

By the same token, overzealous followers of religion might love nothing more than to knock science down and replace it with religion. That's why science and religion need to be kept a safe, supervised distance away from one another.

From my experiences scientists generally make terrible theologians, and theologians generally make terrible scientists.

So, science limits the words that scientists can use. The Theory of Gravity should not imply that there's some internal dispute among scientists about the existence or presence of gravity. Rather, it simply means that the description of gravity embodied in its theory cannot be observed, measured, and validated over the full spectrum of prospects available within a universe we are still trying to understand. Any theory considered validated and thereby accepted by the scientific community is still a theory, even though it's been adequately demonstrated to hold true over some defined set of boundary conditions or portion of the full spectrum of possibilities. The theory can't be tested over all possible conditions in the universe for the simple reason that it's just not possible to do so.

And so is the case for all scientific theories, like relativity, quantum mechanics, and evolution.

Let's cover a couple more quick examples. Science hopefully will find traces of life elsewhere in our universe. Given the sheer numbers of prospective worlds out there, in fact, it's beginning to appear more likely that we will measure the telltale signs of life in the not-too-terribly distant future. Nonetheless, science cannot state that life does not exist outside our solar system because science can't prove the absence of something. Nor does science have the capability to individually examine the fantastically large number of places to look in the universe and state that life does not exist elsewhere. The best science can do is tell us that we have not yet observed any indication of life elsewhere.

Take this one step further. Our observational and measurement skills and technologies aren't perfect, either. For the most part, our technology today remains somewhat immature. This is evidenced by the fact that technology is changing very rapidly, and no one would want to be caught using a smartphone that's more than a year, or at most two years, old. Look at our computers. Sure, they've improved quite a bit, but we continue to rely increasingly on technology that still locks up and fails with frustrating frequency. (At least that's my impression at the moment as a PC user. Apple owners may be feeling smug but, believe me, there's still plenty of room for improvement!)

The point is we might already have looked in precisely the right place to observe life elsewhere in the universe and our technology failed to capture it. This brings us back to the tree falling in the forest. If intelligent life in the universe tried sending us a signal and our sensors were looking in the wrong portion of the spectrum of all possible communications options, would it be useful information to us? Unfortunately, the answer is no. We are forever confined to observing the universe through a perception-of-perceptions process that becomes increasingly complex as we introduce additional layers of hardware, software, and automated analysis technology to the already potentially deceptive layers of perception and interpretation already in place.

Let me provide one more example. Ten years ago science looked out into the universe and determined that the summed mass-energy of our universe is only 4 percent of what we consider it to be today. Only 4 percent! In just a few years, scientists raised their estimates by a factor of twenty-five! (See Figure 7.) If this had been an exam, science would have scored a grade about 50 percentage points lower than an "F."

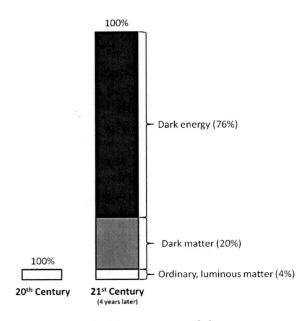

Figure 7. Mass-energy of the universe: 20th Century versus 21st Century.

Now, if we were to go back to the 1990s, scientists might smugly have asserted that we had a good handle on the summed mass-energy of the universe and they still would have been wrong. In fact, the only thing science could have rightly asserted was that our measurements indicated a summed mass-energy of some value. If you or I had pressed science about the prospect of a higher value, the only thing science should have provided was a simple statement along the lines of, "our observations do not indicate a higher value at this time given the methods we used and the current state of our technology." A blink of an eye later and we have more mass-energy than we can shake a stick at.

So, let's introduce an extraneous photon into our experimental apparatus. Our photon is neither shy nor particularly outgoing. Is it possible that we could fail to detect the photon? Of course, the answer is yes.

But, how could this be? First of all, photons are streaming around us all the time, and generally speaking, we are completely unaware of their presence. We even buy products to reduce the observation of unwanted photons, such as sunglasses with polarized lenses. The military routinely exploits a method to avoid, or at least minimize, the eavesdropping on its communications, as yet another example, through the use of laser-based communications systems.

Photons comprising military communications can stream in all directions around a battlefield, but unless you place your receiver (sensor) in just the right place spatially, spectrally, and temporally, there's just nothing equivalent to the sound of a falling tree to hear. But, beyond these arguments is something fundamental about photons and quantum mechanics. There is nothing in the nature of photons and quantum mechanics that requires photons to be observed. Recall that the photon exists as a probability wave before it collapses under the proper conditions to a photon of precisely the characteristics we measure.

We all think of photons in the usual ways. It's the "carrier" of the electromagnetic force. It exhibits wave-particle duality. Scientists refer to it as a particle with a rest mass of zero. However, when you really get down to brass tacks, the photon is a mathematical construct described by its wave equation. Nothing more, nothing less. A photon simply possesses a vanishingly small quantity (hence, the name "quantum") of electromagnetic energy. Its spatial dimensions, like those of many other particles, are poorly defined, and its boundaries are poorly understood.

Photons—as well as electrons, neutrinos, and quarks, and the other "carriers" of force known as the "gauge" bosons—have such diminutive stature that they are considered to occupy only points in space, as opposed to occupying measurable volumes of space. (String Theory uses small vibrating strings to represent these phenomena instead of points.) There's nothing magical about these particles other than the mythical particle status that we confer upon them. In the end, there exist only energy and matter. Einstein's famous equation, $E = mc^2$, shows us the equivalency between them. We also know that energy transfer can only be accomplished by the four fundamental forces.

Let's delve more deeply for a moment into the wave/particle duality of photons. If we try to observe the wave nature of photons, then we can set up a double-slit experiment and observe the beautiful interference pattern produced by our photons—really just quanta of electromagnetism, but only electromagnetism, nonetheless. In this arrangement, we've allowed the "photons" to travel unimpeded to the projection screen. However, we can also collapse the interference pattern by taking a measurement of the photons upstream of the screen.

As we perform our measurements, the "photon" collapses down from its indeterminate probability wave state to a single possibility, and we then measure some particular property of the photon.

We know the "photon" only collapses down to its determinate state when it's measured. Of this, quantum mechanics is quite sure. The phenomena we observe in a sense only come into existence as a consequence of the observer's actions. We never, ever go to measure a photon and observe all the possible states it could take. Instead, we measure only a single possibility and we consider it the reality of this particular photon. Yet, a moment later, after our measurements cease, the quantum of electromagnetic energy returns to its indeterminate, all-possibilities state.

In essence, the wave/particle duality is not really an accurate portrayal of the shape-shifting "photon." A more accurate portrayal goes like this: The indeterminate probability wave describing the photon can't be detected or measured, because anything we could possibly do would disrupt its indeterminate state and knock it into a condition we could then measure.

It's a terribly frustrating conundrum. We can only measure the photon if we perturb it, and we must perturb it in order to measure it.

Is it really possible, then, that other indeterminate states' probability waves "exist" around us that we don't detect as the presence of quanta of energy? In fact, is it possible that we are positively immersed in a sea of undetectable quanta? Well, let me answer this question with another. Is it possible that, only ten years ago, we failed to detect the presence of 96 percent of the universe's summed mass-energy through our measurements? As embarrassing as it might be for all scientists everywhere, me included, I'm pretty sure that I have little choice but to answer that, yes, it is possible.

So, is it really possible that we're immersed in a sea of uncollapsed probability waves and that there's nothing we can do to detect or measure them? That's a great question, but unfortunately the answer is yes, there's nothing we can do. Darn! Let me explain.

Chapter 8
The Heartbreak of Gravity

We know by this point in the book that there are only four fundamental forces. Two of these operate over such small distances that they're not useful in measurements involving anything but the tiniest of distances. These two forces, strong and weak, act only on the scale of an atom's nucleus, which we've calculated to be just a teeny-weenie portion of an atom's radius, while the radius of an atom is on the order of an angstrom. So, the strong force and the weak force are of little use to us for any type of measurement.

We just finished discussing that the quanta of electromagnetic energy representing the prospective phantom photons around us may be unobservable, indeterminate states that elude our detection. So, depending on electromagnetism to measure them is out.

That leaves us with only one remaining force upon which to base our measurement: gravity, the force that stands somewhat apart from the other forces.

There's some good news with respect to gravity. Recall that Einstein showed the equivalency between mass and energy, and he showed us that mass curves space. Einstein went on to demonstrate that this curved space is actually the basis for gravity. Unlike Newtonian physics, in which the concept of gravity was described as an attractive force between two bodies that gets reduced by squaring the distance between their centers of gravity, Einstein's physics presents us with the concept that each object is falling into the gravity well produced by the curving of space around the other's mass. Furthermore, as long as an object is comprised of stuff we call matter, whatever that really is, it curves space and there's gravity.

Eureka! We'll use gravity in our quest to detect the phantom photons.

The illusion that we can use gravity to perform this measurement is compounded by two other simple assertions we can make about gravity. First, as of today, we are unaware of any way to shield gravity from measurement, meaning that nature has no way of hiding gravity from us. Second, regardless of whether an object is comprised of regular matter, antimatter, or the for-now-mysterious dark matter (of which we've heard so much over the last few years), gravity is always attractive and never repulsive. Curvature of space is curvature of space, it seems. So, gravity always possesses an attractive force, since the curvature of space is always inward toward the center of mass or gravity of an object. If, in fact, there exists some basis for an opposite curvature of space, it has yet to be observed or even postulated, for that matter.

So, before we set out to begin our gravity measurements, we need to learn more about gravity. This is where the bad news creeps in.

Gravity is notoriously difficult to measure. In fact, of all the measurements that are made in an attempt to measure the basic properties of our universe, gravity is the most difficult. Just how peculiar gravity is compared to the other forces begins to become brazenly obvious as we try to measure it and observe it.

I mentioned the concept of gravity that Newton developed earlier, but let me introduce the rather simple and straightforward equation that he provided to us:

$$G = k \, \frac{m_1 \times m_2}{r^2}$$

This is a perfectly good equation to use as long as we recall that, according to Einstein's equations, this is really only an approximation that is useful over a particular portion of the complete range of circumstances possible in nature. For our particular set of circumstances, it's a great equation to use.

Let's define the parts of the equation to make certain that the concepts contained within it are clear:

◊ G is the amount of gravity that we're calculating between two objects

◊ k is the universal constant of gravitation, one of only a handful of parameters which describe the fundamental nature of our universe.

◊ m_1 is the mass of the first object, let's say our planet Earth.

◊ m_2 is the mass of the second object; let's say a 100-kilogram person standing on the surface of Earth at sea level.

◊ r is the the distance between their centers of gravity, which for the situation I just described is about 6,400 kilometers.

Henry Cavendish was the first scientist to attempt a measurement of the universal gravitational constant, k, at the end of the 1700s. Since then, countless refinements have been made to the experiment, which has been performed count-less times. Over the few hundred years since Cavendish first performed the experiment, the value for the universal gravitational constant has been refined to today's accepted value of 6.67428×10^{-11} m³kg⁻¹s⁻².

There's no way of knowing the actual value of the universal gravitational constant; all we know how to do today is to measure it with improved precision, but we can't compare the value to an actual value because the actual value is unknown. The reason why the actual value is unknown is that no one has ever successfully developed a means by which to calculate it.

We've talked about building models from the most fundamental principles (the first principles) in an attempt to understand the nature of the universe. In many cases, we can build a model that allows us to calculate some value that we call our prediction, and then we can compare it to what the universe is telling us to determine how well our prediction stacks up. However, when it comes to something as fundamental as the universal gravitational constant, the best first principles-based prediction that scientists have made is off by forty orders of magnitude! In case you're not a science type and the drama of the last statement was completely lost on you, science's best prediction of the universal gravitational constant has the decimal point off by forty places! Hopefully, you're now stunned.

Once again, if we place this level of performance in the form of an exam grade, it's equivalent to staying in bed and not even showing up for the test. Science gets a grade of zero to forty decimal places.

Here's the point: The value of the universal gravitational constant is in itself another paradox about the nature of the universe that emerges from the current state of our scientific understanding.

So, the gravity card we were hoping to play isn't turning out to be what we might have hoped.

Essentially, our challenges in using gravity strongly relate to our interest in using it as the basis of a local measurement, one made in our nearby vicinity. Let's explore this idea and see where it's most useful.

So, again, let's compare and contrast the four fundamental forces from the perspective of the measurement we would like to make:

◊ The strong force keeps protons together in the nucleus of an atom and operates over a short distance. Scientists have measured the strong force as possessing a strength of 10^{38} compared to gravity. This force is very strong.

◊ The weak force keeps neutrons together in the nucleus of an atom and operates over a very short distance. Scientists have measured the weak force as possessing a strength of 10^{25} compared to gravity. This force is very strong but not nearly as strong as the strong force.

◊ The electromagnetic force is the basis for all forms of electromagnetic energy, including radio waves, microwaves, light, x-rays, and so on. This force operates over very large distances but drops off quickly with the square of the distance. Scientists have measured the electromagnetic force as possessing a comparative strength of 10^{36}. This force is stronger than the weak force but weaker than the strong force.

◊ The gravitic force is the force arising from an object's mass and its curvature of space. The gravitic force manifests itself as an attractive force between any and all objects or phenomena possessing mass. Gravity operates over very large distances but drops off quickly as the square of the distance. For our comparison, gravity has been assigned a comparative strength of 1. Gravity is puny compared to the other three forces. (Keep that paper clip exercise in mind whenever you need to think of some way to compare the strengths of gravity and electromagnetism.)

Let's also think for a moment about what it takes to hide each of these forces from our probing sensors. All you need to shield the strong and weak forces is just a little bit of distance (space) and the forces become undetectable.

We can shield out electromagnetism in a few ways, depending upon the portion of the spectrum that we're considering. We could use electrically conductive plates or screens comprised of metals with certain properties to stop electromagnetism in its tracks. This is precisely what manufacturers of microwave ovens do in order to protect consumers from the harmful effects of microwaves on living things.

Of course, if we're talking about visible light, we can shield it using opaque materials that reflect or absorb the light, thereby preventing the light from penetrating where it's not wanted. Putting distance between a source of electromagnetism and our sensor is yet another useful method for shielding. We can take advantage of the fact that electromagnetism drops off with the square of the distance to diminish the magnitude or brightness of the electromagnetic signal at our sensor as the electromagnetic energy expands into a greater volume of space.

Now, once again, we come to gravity: our remaining, peculiar force. The only method available to us to shield gravity is to take advantage of the fact that gravity, too, drops off in magnitude with the square of the distance between a particular mass and the apparatus that we use to measure it. Science currently knows of no method to shield our sensors from the curvature of space that is gravity. Like everything else in our world, this fact represents a double-edged sword that works at times to our benefit and at other times feeds our frustrations.

So, if the strength of gravity drops off with the square of the distance, can't we flip this fact around to help us? Why not look at the glass as half-full instead of half-empty? If we want to increase the strength of gravity that we measure, why don't we get really close and intimate with the objects we want to measure. The fact that gravity strengthens as we approach our object or phenomenon should make it easier for us to measure, right?

Wrong, very wrong. In fact, the opposite is true. The reason is quite simple. If we try to perform table-top laboratory experiments of gravity, the amount of gravity we would be trying to measure is immeasurably small, and we can't shield out the contributions to gravity from the objects that surround it. So, things like the gravity detector itself, our table, the building around us, the air we're breathing, the planet beneath our feet, and our very own presence all conspire to make the measurement impractical. The laboratory setups and sensors for performing gravity measurements can be quite large.

Trying to perform a gravity measurement in a lab setting is not terribly unlike trying to measure the gravitic force between two objects, like marbles immersed in a swimming pool. No matter how we try to alter our laboratory setup, we can't deviate sufficiently from the "measurement in a pool" analogy.

So, what might a good gravity lab look like? Well, it would look something like this: big masses separated from one another by empty space.

The bottom line is that if we want accurate gravity measurements, we have little choice but to look out into the depths of space and observe the behavior of really massive things like stars and galaxies using electromagnetic-based observation schemes. We have no effective way to precisely detect or observe small amounts of matter and the feeble gravity signatures they produce at close range.

This leads to two very simple hypotheses:

1. If there were large amounts of non-luminous matter in the universe, we ought to be able to detect their gravity signatures indirectly through the behavior of distant massive objects observed via electromagnetism.

2. If there were large amounts of non-luminous matter surrounding us in our vicinity, we have no effective means of detecting it.

We're going to focus on the latter hypothesis first. If the local matter is non-luminous, meaning that it does not or will not provide us with an indication of its existence through electromagnetism, science has no way to verify its existence; science, as a result, would place it on the shelf reserved for religious beliefs. Yet, science is completely unequipped to prove the absence of anything, so perhaps—just perhaps—science is being a bit hasty and should slow down for a moment until it thinks this dilemma through a bit more.

Let's return to the former hypothesis now. As ludicrous as it sounds, science has firmly established over the short span of the last decade the existence of dark matter. Non-luminous dark matter cannot be detected through direct electromagnetic-based observation and instead is detected through indirect schemes that are based on observing the motion of luminous matter. In fact, there's a boatload of dark matter out there. Based on the most recent measurements and interpretations at the time that I was writing this book, for every morsel of ordinary luminous matter there appears to be five morsels of dark matter! In other words, in terms of the matter science now observes "directly" through its electromagnetic properties and "indirectly" through its mass and gravitic properties, only one-sixth (about 17 percent) of matter is ordinary, luminous matter and five-sixths (about 83 percent) is non-luminous dark matter. Needless to say, for all science knows there may even be more matter awaiting discovery, but for now this is plenty of matter for us to digest. Indeed, at the moment, it's far more matter than science currently knows how to digest.

Of course, the prospect that we just didn't notice 83 percent of the matter surrounding us could complicate our attempts to understand and calculate the universal gravitational constant.

Remember, this process I'm proposing doesn't permit us to assert a new scientific truth. It only permits us to identify plausible prospects that science can act upon, ignore or discard. These hypotheses may or may not be true, and mathematicians and physicists will need to weigh in with new models, calculations, experiments, and observations.

The history of science and philosophy is fraught with the notion that there's something special about our place in the universe. Early scientists and philosophers believed that the Earth was the center of the universe. With great turmoil, that idea eventually gave way to the notion that our sun was the center of the universe. Again, with great turmoil, these geocentric and heliocentric views of the universe gave way to the concepts inherent in Einstein's relativity, in which we now know that there is nothing privileged about our particular inertial framework or region of space. The consequence is that we are only beginning to understand that there may be nothing special about us, our world, or the particular part of the universe we occupy. The impact of such advancements on human culture is huge, as the image we formed of ourselves over our collective human heritage gets replaced with a new one in which our apparent importance in the overall scheme of things is diminished.

Along these lines, many human cultures and institutions have believed for a very long time that life is unique to our planet and yet, with each passing day, we learn of more and more potentially life-friendly worlds even in our general vicinity of the Milky Way Galaxy that possess the prospect of harboring life.

Yet another example of something very similar to geocentrism and heliocentrism prevails in science even today. The perceptions we derive from some of our scientific measurements would seem to indicate that we are at the center of what has been called the "Big Bang," that we are at the center of a rapidly expanding universe, and that all objects in our universe are moving away from us in what is suggestive of Earth having a really bad case of body odor. At the same time, our instruments tell us that the lingering echoes of the Big Bang are uniform in every direction around us to the improving precision that we achieve with our technology.

I seem to be seeing another manifestation of this same sort of thinking but this time somewhat in reverse. This time, scientists perceive the presence of dark matter far from us, apparently thinking that these faraway places are somehow special. My experiences and the common sense I've developed over my lifetime would suggest to me that it's far more plausible that there's nothing special about those particular faraway regions of space whatsoever. To the best of my knowledge and understanding, each and every single attempt that humanity has made to impose special stature on certain regions of space, be it space over here or space over there, over other regions of space have fallen flatly on their faces. I consider it far more likely in this situation that what we are learning about "space over there" probably holds true for "space over here" as well. Through our ongoing investigation of distant space and the dark matter phenomenon, we are likely learning quite a bit more about ourselves, too.

Perhaps there's a plausible explanation for these phenomena that accounts for the special nature that our measurement seems to indicate. Perhaps, indeed.

Let's return for a moment to the discussion we had a little while earlier about the futility of performing local gravity measurements. Let's revisit in particular our analogy involving gravity measurements of objects in a swimming pool filled with water. If we replaced the water with something denser like mercury while holding steady the mass of the marbles we are using to perform our experiment, the increased mass of the mercury over that of the water it replaced makes our measurement even more difficult.

Could this implication be among the possible explanations that account for our inability to calculate something as fundamental as the universal gravitational constant? It would seem pretty reasonable to me that it could, since the presence and distribution of matter both locally and over great distances is at the crux of this calculation. Please don't think I mean that the mass of the Earth is different than the value we normally accept. Remember that any attempt to measure the mass of the Earth already includes whatever undetectable matter might be lurking within and around it. This directly unobservable stuff is built right in to the observations and measurements, so we'd never notice it as a separate concern.

We used our abstract reasoning skills along with scientific evidence and the benefit of hindsight in an attempt to reduce our hypothesis that there's such a thing as electromagnetically undetectable matter and that it exists around us, to the absurd, and discovered instead that the conclusion we reached demanded a seemingly absurd condition to be true. However, the absurd phenomenon we predict is indeed observed and considered by science to be a paradox. There's a whole bunch more dark matter than luminous matter.

So, let's return to our single-photon, double-slit experiment and see if we can hear what it might be trying to tell us in light of the exercise we just completed.

By way of review, we allow only one photon to enter our experiment at a time and we release the next photon only after the previous photon passes through the slits and hits the projection screen. We record the pattern of the photon strikes over time and we see an interference pattern. Perhaps this phenomenon doesn't appear as cryptic as it did earlier in the book when I first introduced it to the reader.

I'm now going to share my interpretation with you but I'm going to provide a play-by-play analysis in reverse. In other words, I'll begin with the observation and work upstream, back to the light source.

First, as I mentioned, we observe the interference pattern as we integrate our observations over time. This tells us that, indeed, waves are interacting. The other crucial piece of information is that we observed (detected and measured) a photon strike possessing a single set of quantum characteristics, so the last thing our electromagnetic quantum did was to take on a determinate set of characteristics as we observed it striking the screen.

Immediately prior to this event, the quantum of electromagnetic energy "existed" as an indeterminate probability wave described by its wave function that possessed all of the prospects that include the single determinate value that the photon will soon be taking. I placed the quotations around the word *existed* because this violates the traditional quantum mechanical definition of existence. In fact, as we'll learn soon, the pre-Socratic Greek philosophers (the Atomists) struggled with the concept of existence and non-existence nearly two and half millennia ago. In many ways, the concepts of existence and non-existence embodied in the philosophies and writings of the Atomists remain crucial to our understanding of the universe, and these issues remain unresolved. As I also hope to share with you, this lack of resolution about the aforementioned concepts, and the accompanying discussion of atoms and void, directly impinge upon the emergence of the paradoxes that we're now trying to overcome.

Again, our energy quantum is now in an indeterminate state described by its quantum probability wave function. At this point the quantum is travelling as a wave and it has been traveling through space as a probability wave since the time it was emitted from its source. It travelled as a wave through some region of space described by Maxwell's equations with its electric and magnetic fields undulating in orthogonal planes, and then it passed through the slits in this indeterminate state and continued until the screen interrupted its travel and we performed a measurement. During this time the probability wave that will soon become the basis of our observation had ample time to interact with the indeterminate quanta represented by other probability waves that we don't know anything about (and never will) and an interference pattern was the consequence.

Well, the part about never knowing anything about the other probability waves can't really be true now, can it? After all, the undetectable probability waves will ultimately betray their presence by contributing to the formation of the interference pattern. So, similar to dark matter, we can't directly observe these particular phenomena because they're not directly detectable, but we sure can observe the consequences of their existence. Paradox, what paradox?

Again, throughout this process we remained loyal to scientific observation, but we certainly departed from traditional scientific interpretation and doctrine.

Let's now turn the situation back around to where we started on our discussion of this whole double-slit experiment phenomenon.

If we seek to observe the wave nature of photons, we do so through observation of the interference pattern produced by the double-slit experiment. If we attempt to perform a measurement of the photons along their path between the light source and the projector screen, then the interference pattern associated with the photons' wave natures collapses and gets replaced with a different pattern, and we observe the particle nature of photons. If, instead we perform a single-photon, double-slit experiment, we still observe an interference pattern so long as we don't attempt to perform a measurement of the photons along their path.

Does the phenomenon seem so cryptic and mysterious now, I ask you?

I realize I have many more phenomena to explain to round out the picture, but I can only proceed with you one step at a time. My job is certainly not yet complete, but it has begun.

We'll be going through additional exercises like this, one by one, so I can demonstrate to you the possibility that many of today's scientific paradoxes are related to each other. The perception that there are numerous paradoxes emerging from our current scientific understanding may be in itself an illusion. Eliminate one or more of the paradoxes and its cousins may come tumbling down like a row of dominoes. A far different and simpler universe may be awaiting us.

Chapter 9

And Now for a Completely Different Variant of the Double-Slit Experiment

Now it's time to return to and encounter the next variant of the double-slit experiment. Recall the significance that Dr. Richard Feynman placed on the double-slit experiment and the prospect for understanding the universe that he believed would result from getting our arms fully around the perceptions we gather from it.

This time we're going to place two separate double-slit experiments on our lab bench. Everything is identical to our previous single experimental setup, except the light source. In all other regards the experimental setup is identical to our previous one; furthermore, the two experimental setups sitting on our lab bench are identical to one another.

Here's what we're going to do with our light source that makes it different from what we did before. We're going to use a single source of electromagnetic radiation (a fancy name for a lamp) and we're going to allow the radiation we produce to strike a crystal. It's a particular type of crystal, to be sure, but it's just a crystal any way you cut it.

A crystal consists of a periodic arrangement of a basic molecular structure, called a unit cell. All of the unit cells are lined up facing the same direction, like a bunch of soldiers all lined up for inspection. There are few, if any, missing soldiers in our high quality crystal. As we mentioned earlier, materials possessing this crystalline state have certain properties that can be very useful in the proper applications.

Certain crystalline materials have particular properties we want to exploit. We'll use a special crystal to provide us with two photons for each photon that strikes it. Since the energy involved in this reaction needs to be conserved, each of the resulting photons will possess half of the energy of the photon that spawned them.

Like all families, this family of photons possesses certain common features that are strongly related. In this case, our "momma" photon spawned a set of identical twins. As soon as the offspring are produced, they are identical to each other in every imaginable way. And the twins look a lot like their mom but not identical. Since the amount of energy in the reaction needs to be conserved, each offspring photon possesses half the energy of its mother and a wavelength that has twice the value of its mother, so the combined energy of the twins is equal to the energy possessed by the momma photon.

Quantum mechanics informs us that these two quanta of electromagnetism are identical, but that pesky aspect of measurement gets in our way, again and again. For, if we attempt to measure the photons, the rules of quantum mechanics require that the two photons cannot each possess all of the same attributes. Their complete quantum descriptions must not be identical, so, not unlike twins in the macroscopic animal kingdom, we will indeed find a way to distinguish one sibling from its twin if we examine it carefully by performing a measurement.

The big difference, however, between twins in the quantum world and those in the animal kingdom is that we only need to observe or measure one photon of the twin pair to know how to distinguish one from the other. That's because of a peculiar aspect of the intimate quantum relationship that exists between these produced twin photons called "entanglement."

When scientists really want to wow an audience and view a sea of stunned, perplexed faces, entanglement is the topic they commonly choose. As we proceed with our new double-slit experiment, the reasons for this will become more evident.

Let's finish defining our experiment. We turn on our lamp to produce electromagnetic radiation in the form of photons and allow them to strike our crystal. A reaction within the crystal produces a pair of twin daughter photons from each incoming photon. We allow one daughter photon to enter each experimental apparatus. Each photon follows the now familiar steps of passing through the slits and striking the screen where we perform our observations and take our measurements. So, for each momma photon we produce, we're going to collect two sets of measurements: one for each of the daughter photons.

We're now set and ready to go! We turn on the apparatus and begin our observations. With a deep sense of comfort, we notice the now-familiar, expected interference pattern. And, just like before, if we attempt any measurement of a photon that could provide us with any insight as to the particular path through the slits that this particular photon took on its way to the screen, the wave interference pattern collapses. Everything is working as it should. We feel a sense of harmony with the universe. Life is, indeed, good.

After a while of collecting measurements, we decide to review and interpret our data. During this particular experiment, though, we had decided to collect some additional information about each photon we measure. We decided to collect information about a certain quantum attribute of photons called "spin." Spin can take on only one of two values for the information we record is binary. This means that any measurement is comprised of one value out of only two possible choices. We could record it as either "this" or "that;" we could use "zero" or "one;" or we could use "up" or "down," as long as we maintain a consistent system. Since the field of quantum mechanics has selected the "up" and "down" nomenclature, we will just go with that one.

We scrutinize our observations and we find it reassuring that each observation of a photon from one apparatus has a matching entry from the other apparatus. That's good. Since we are now using pairs of photons, we should be acquiring matched pairs of observations.

As we begin to analyze the data that we recorded, we noticed something peculiar, though. Each pair of observations with respect to spin looks like "up/down" or "down/up." Each pair of data consists of one "up" and one "down." We don't see a single pair of "ups" or "downs" anywhere in the data.

Could this really be correct? Hopefully, by now you've learned that if the observation is peculiar, it's probably an accurate reflection of what science observes about the universe. So, yes, it's correct and this is what entanglement is all about. The ramifications of entanglement are nothing short of shocking, to say the least, as I'll explain. There are just so many aspects of entanglement that are shocking; they each shake the foundations of how we normally interpret our universe.

This particular form of entanglement is called antiparallel, since the daughters take opposite quantum spin values. It turns out that there are other forms of entanglement in which the twin daughter quanta take on identical values, or what we could call "parallel" entanglement.

So, we've identified a sort of deep familial dependence between this pair of simultaneously spawned twin quanta. Let's go ahead and explore more about entanglement.

Back to the lab! We turn on our lamp, producing electromagnetic quanta that strike our crystal. Entangled quanta pairs stream out of our crystal, one of each pair heading toward one set of slits, and the other of the pair heading toward the other set of slits. We quickly engage our phase shifting filters from one of the earlier variants of the double-slit experiment, and we place them over one set of slits so we can observe the left photon and determine through which slit our left photon passes on its way to the screen. Just like before, the wave interference pattern on the left screen collapses and gets replaced by the particle pattern. But so did the pattern on the right!

The measurement of the photon traveling through the experimental apparatus on the left caused both interference patterns to collapse! Hmmm ... is that really as peculiar as it might seem at first?

Let's employ our understanding from the previous version of the double-slit experiment in an attempt to interpret what we just observed.

A photon emerges from our lamp in the form of an indeterminate probability wave and we make no attempt whatsoever to intervene in any way. This photon enters our crystal, and two entangled daughter photons are spawned. Even though we are now the proud godparents of an entangled pair of bouncing baby photons, each possessing half the energy as their mom, we successfully resist the urge to take any pictures of the twins to use as a wallpaper background on our phones, so no measurements have been taken and no observations of any sort have been performed.

As a result, our newborn quanta remain in an indeterminate state until one of them encounters the phase-shifting filter in the set up on the left as the first step in our measurement. Then the quantum is forced to leave its indeterminate state of daydreaming and enter a definite state, so we observe, once again, a very definite set of quantum state values. Now, this is no "ordinary" set of twins; this is a pair of entangled quanta. What scientists observe over and over again without exception is that, when one quanta in an entangled pair collapses down to a determinant state, they both must do so together.

So, even though we chose the left quanta as our favorite godchild and were ignoring our godchild on the right, the quanta on the right could no longer sustain its indeterminate state, and its state collapsed down to a definite, measurable photon whose spin value is the opposite of its twin sibling.

If we then remove the phase shifting filter on the left, each interference pattern re-emerges on its respective screen. Scientists have attempted many variations of this particular variant of the experiment and, like all of the other variants, the results are completely consistent. You can try different measurement techniques till the cows come home, and the results don't change. Of course, it just doesn't matter whether you attempt to perform your measurements on the left apparatus or the apparatus on the right.

Hopefully, the experiments we've discussed with entangled quanta make sense to you and you're okay with moving ahead; there's more and it's a whole new level of peculiar.

As we discussed, if I measure one quanta of this entangled pair, its indeterminate state collapses to a definite observable state, resulting in the collapse of its sibling's indeterminate state to a set of definite observable quantum states, each with an opposite quantum spin value. You might wonder: How long does it take between the collapse of one quanta's indeterminate state and that of its sibling?

Let's think about this carefully for a little while before we guess anything hastily.

As you are likely familiar, if you've chosen this book to read, there was a scientist named Albert Einstein who was troubled by the paradoxes about our universe emerging from the science of his time at the start of the twentieth century. He devised his theories of relativity that provided an unprecedented window into the nature of the universe. Integral to his theories of relativity was a feature of the universe he deduced from "thought experiments," or *Gedanken* experiments, for which he was famous. Before Einstein's relativity, there was a debate among scientists as to whether or not the speed of light had a constant value for all observers, regardless of where the observer was located and the ways in which the observer might be traveling with respect to the phenomena he observed. He demonstrated conclusively that the speed of light is constant for all observers in fixed inertial reference frames and the impact upon human culture, both the scientific and non-scientific communities, has been staggering. In all likelihood, his name will be a household word for a very long time.

If ever anyone wanted to establish a "speed limit" fundamental to our universe, the obvious candidate would be the speed of light—represented by the symbol "c" and which has a value of about 300,000,000 meters per second. Einstein believed that the speed of light—or, more accurately, "the speed of the propagation of electromagnetic waves through free space"—established the ultimate speed at which information could be conveyed. Although there are some candidate phenomena that might travel at speeds greater than the speed of light, he believed that information exchange could not occur at a speed beyond the speed of light. Again, this amount of information is probably sufficient for our purposes. I'll provide more details later.

Back to entanglement and the question of how quickly the collapse of one quanta's indeterminate probability field affects the resulting collapse of that of its sibling. The side discussion about Einstein and the speed of light might lead a rational person to predict that one photon would try to signal the other about the collapse of its indeterminate state and that information would be transmitted at the speed of light, right? And, furthermore, a rational person would also predict that, based on the speed of light, the time it would take to report its condition would depend on the spatial separation of the two entangled photons, right? I mean it, right?

Scientists have spent a great deal of time and effort on measuring the time between the collapses of two entangled photons. They have honed their technology and experimental setups to improve the precision of these measurements. Here's a summary of what they measure: The time between the collapse of one photon's indeterminate state and that of its sibling happens faster than the speed of light. In fact, it happens much, much, much faster than the speed of light (with today's best estimate suggesting that the time between the collapses represents no less than 10,000 times the speed of light, and perhaps far, far faster than that). Hold on, please; it's about to get really, really strange.

Furthermore, no matter how the experiment is performed, the timing between the sibling's collapses doesn't vary with distance!

In other words, you can separate your two experimental setups any way you choose, and there's no change in how quickly the paired collapse occurs. You can place each setup 10 or 100 or a billion miles away from your crystal and send the indeterminate quanta pair in opposite directions until they reach their respective setups, measure one quanta, and the other quanta collapses, perhaps instantaneously, to its opposite determinate spin state.

Let me be very clear about this: Scientists are absolutely unable to rule out the possibility that the phenomenon occurs instantaneously because, no matter how accurate of a clock they construct, the time between entangled collapses is immeasurably small. In case I'm not making myself really clear, it is possible that the phenomena occurs without the passage of any time and it appears that this would be the case even if the entangled quanta were on opposite ends of the universe!

Perhaps you are asking yourself if this could possibly get any stranger. Well, it does. It even gets really stranger. To see just how bizarre the situation gets, we need to revise our experimental setup once again.

The basic experimental setup doesn't really change from this point onward as we examine the few remaining variants that I've chosen to include in my book.

For this variant, we'll keep everything the same except for one of the two double-slit panels and its associated projection screen. We'll keep the lamp, the crystal, and the left portion of the experimental setup as it is. Then, we'll relocate the right double-slit panel and screen down the road a good distance. It could be any distance, but let's set it at ten miles. Furthermore, since we don't want to allow air molecules to affect our experiment, we will send the photons to the ten-mile-distant right side of the setup through a pipe from which we've drawn a vacuum.

Now, we turn on the lamp, produce our entangled pair of quanta, send one quanta of each pair to the nearby left side of the setup, and simultaneously send the sibling quantum down the ten-mile-long pipe to the right side of the setup. We then measure the photons on the left. The quanta on the left collapse to measureable, determinate quantum states as do their siblings on the right side of the experiment, and the interference patterns disappear.

Sigh. Not a single surprise, since this is the result we had come to expect—strange as it is.

Let's contemplate what's going on in this experiment before we take any further steps. From each entangled pair of spawned quanta, one travels through the slits on the left and hits the screen very soon after its production, since the indeterminate packet of electromagnetic energy is traveling at the speed of light and the left screen is very close to the crystal. The photon on the right, which we sent down the pipe, has to travel ten miles further before it reaches the slits and screen at its end. So, from each entangled quanta pair, one hits the left screen at a certain time and the sibling quanta reaches the right-hand screen later, since it's traveling ten miles further at the speed of light through a vacuum.

As we measure the photon on the left and its probability wave collapses to a determinate state, the sibling photon traveling to the right "instantly" collapses to its related but opposite determinate spin state well before the photon on the right ever reaches the slits or the screen. In fact, if the transformation is instant, then the transformation is complete before the photon reaches its destination, since the photon didn't even have the opportunity to travel down the pipe before the entangled pair collapsed to their determinate states.

Perhaps this much all makes sense so far, because we're about to stand common sense and experience on their heads.

We're now going to introduce a tiny but very significant change. Instead of observing and measuring the photon on the left, we're going to get on our bicycles and ride down to the end station of the experiment on the right. Now, we're going to measure the photons on the right and see if we observe something peculiar. Remember that by the time the right side photon reached our measurement apparatus on the right side, the photon on the left had reached its destination earlier. So what could possibly happen?

We turn on the flow of entangled photon pairs; we wait for the right-side photon to reach our measurement station; we perform a measurement on our right-side photon; and we discover that the interference pattern and the indeterminate probability wave of the photon on the left both collapsed. The wave pattern on the left collapsed retroactively! Once again, I suggest the proper response is, "Huh?"

The measurements scientists record from this experiment unequivocally indicate that the photon state on the left collapsed retroactively in time in order to ensure that the collapsed states of the two sibling photons were harmonized and consistent. In fact, some results even suggest that the photons will collapse to their respective determinate states as soon as the intention to perform a measurement is established.

From one moment to the next, our prevailing concepts of space and time have been threatened and replaced by the prospect of some sort of hoax, some sort of illusion. We're confronted by a difficult choice that demands that we discard our long-held notions of time and space or the careful and repeated observations performed by some of the most qualified, eminent scientists of the last 100 years. But the results are clear and they provide an example that causality, or at least our concept of causality, has been violated. All it takes is one violation to tell us that one or more parts of our interpretation of the universe are very wrong.

Einstein himself referred to this phenomenon disbelievingly as "spooky action at a distance," and refused to accept it. The phenomenon has become known as the EPR Paradox, named after a paper authored by Albert Einstein, Boris Podolsky, and Nathan Rosen. In fact, the EPR Paradox was a central theme of the Copenhagen Interpretation of Quantum Mechanics.

It may indeed seem like we're in a cosmic pickle, but it appears as though there may be a way out of it. Quite frankly, this is going to take one really good explanation to extricate us from this pickle and banish the paradox.

This seems like a particularly good time to bring up another key element of Einstein's relativity. You're probably already familiar with it. In Einstein's theory, the three dimensions of space and the dimension of time combine into a single framework, which Einstein referred to as the space-time continuum. In essence, Einstein said that the separateness of space and time are illusory and that they need to be combined into a single four-dimensional construct called space-time. Depending upon the way in which observers occupying one perspective, or inertial framework, view objects or phenomena that occupy a second, different inertial framework, they experience changes in time and space or, more accurately, space-time, in order for the speed of light to remain constant for any and all observers.

Another way of stating this is that space and time depend upon some aspect of the observer and his or her motion through space. Einstein's theory—and the myriad experiments that have been performed to conclusively demonstrate different aspects of his theories—clearly demonstrate that time for each observer runs on a clock unique to each observer. We each carry our own unique clock in which time unfolds uniquely. Under your usual, normal, everyday conditions on Earth, with typical motions occurring well under the speed of light, this means that each of us is pretty much running on the same time. But under different conditions, this no longer holds true, and time could run at drastically different rates for different observers. Under the appropriate conditions, space-time bends this way and that. Time and space bend in order to make sure that each observer under any possible set of circumstances observes the very same value for the speed of light. Each individual observer carries his or her own unique yardstick by which to measure distance and his or own clock by which to measure time.

Here's another way to sum up these ideas. Time and space, as we normally think of them in our daily lives—or space-time, as Einstein would have it—are not exclusive properties of the universe. They are properties also of the observer. Einstein said it! It says so right in his writings on the subject. Science has confirmed it! It's just not a point for debate, even though it defies our most basic beliefs and understandings about our universe.

I spent most of my life mesmerized by Einstein's relativity, voraciously reading books by Einstein and others who further interpreted his work, and yet the real meaning and consequences of his work had not become apparent until I read Dr. Robert Lanza's statements in *Biocentrism*. It took only a few moments for the consequences of Dr. Lanza's statements to settle in and for the concept that space and time belong in some way to the observer to alter my internal universe in truly profound ways.

I had been waiting for this moment, and there was little doubt that this was the moment for which I had been waiting. The force and truth embodied in this concept rewired certain circuits of my mind, apparently, and I saw a way out of the current paradox. I was no longer tethered to an interpretation of space and time for which no solutions to the paradoxes of the universe could be found. To say that the pieces finally fell into place would be a very strong understatement.

So, our notion of time and space is wrong. Of this there can be no doubt. Of course, the most troubling aspect of this is that nothing meaningful has emerged as a replacement to the concepts of time and space that we're being forced to abandon. That's the really troubling part of this whole thing. In fact, scientists, as well as everyone else who knows about the paradox, realize that they must absolutely abandon this faulty, romantic notion of time and space. But they don't—for two reasons, I suspect. I've already mentioned the first reason—that it becomes a bit impractical to abandon our notion of time and space and replace it with some new notion of time and space that's understandable and useful, when this hasn't yet happened. The other reason, though, is that I suspect it's culturally too difficult for us to really face the simple fact that time and space are not the fundamental components of the universe we presume them to be, based on our lifetime of perceptions, so we choose to ignore this pesky fact and go about our daily lives as though nothing has changed.

Just as human culture continued to embrace Newtonian physics as a sufficient model to serve as the basis of our daily lives, even after the emergence and acceptance of Einstein's relativity, I suspect that any new notion of time and space will continue to have little impact regarding the daily perception of individuals—even if the impact of a new notion of space and time powerfully affects our science, technology, philosophy, and religion.

Einstein's work forced us to abandon long-held cultural beliefs about the apparent differences between matter and energy, as well as the apparent differences between time and space, and these concepts very much served as the ways in which people distinguished between "this" and "that" when perceiving their world. People always seem to want or need a way to divide the grandness of the universe down into smaller, manageable pieces—not unlike the way our minds deal with the abstract notion of quantity, as we learned in our earlier exercise involving the conceptualization of quantity through the counting of imaginary objects.

Some of the ways in which people commonly divide up the world will likely continue. We're familiar with some of the old-fashioned schemes, like light and dark. Our discussion involving whether a falling tree makes a sound illuminated, no pun intended, our perception of sound—that sound is only a very small portion of the acoustic spectrum that our ears can "hear." In the same way, light is only a very small portion of the electromagnetic spectrum that our eyes can "see."

Other old-fashioned ways of dividing the world up include concepts of good versus evil and us versus them. Such schemes can only be vaguely defined and usually expose the nature and extent of human biases. So, are we left with a simple scheme for breaking the universe down into manageable bites that might be useful to us as a starting point in developing a framework to replace our traditional model of space and time? Yes, in fact, there is a simple scheme, and it's among the oldest, old-fashioned concepts possessed by humanity. It involves the philosophical concepts of matter and the void that predate science itself.

To get past the paradox presented by entanglement, we need to take a step back—a really big step back. We need to revisit the dawn of philosophy and understand the very deepest roots of science and the questions posed by philosophers that have remained unanswerable and unanswered. We're going to step back 2,500 years and discover that certain notions like relativity are actually nothing new.

But, before we do that, I have some additional comments to make and some new entries to add to our glossary.

As you are well aware, another way to divide the universe into two pieces is this: living versus inanimate. It might not have been terribly obvious, but the topics we've discussed and the perspectives we've gained through these discussions now allow us to make a simple, objective distinction between the living and the inanimate based on our new understandings of the concepts of observer and observation.

Here are the new items I propose for our glossary:

◊ **Life** – (a) the state of matter that meets all three of the following criteria: (1) possessing self-awareness of internal states, (2) capable of perception, and (3) *incapable* of existing in an *indeterminate* state with an uncollapsed probability field, since the state of matter comprising the living entity is under continuous self-observation; (b) the state arising from a self-observable arrangement of matter resulting from self-awareness of internal states.

◊ **Non-life** – inanimate objects and/or phenomena that meet all three of the following criteria: (1) not possessing self-awareness of internal states, (2) incapable of perception, and (3) capable of existing in an indeterminate state with an uncollapsed probability field when not observed.

So, let's go back to Greece, about 600 years before the Common Era, when philosophers questioned the nature of reality.

Chapter 10

The Age of Philosophers

The mortals lay down and decided well to name two forms (i.e. the flaming light and obscure darkness of night), out of which it is necessary not to make one, and in this they are led astray. (Parmenides)

What is truth? Where is truth? In this section, I'll begin reviewing some key aspects of Eastern and Western philosophy regarding matter, reality, and the universe. The time period that I cover spans roughly between the sixth century BCE to the second century CE. Interestingly, the questions with which humanity has struggled had begun to be asked by the thinkers and philosophers in disparate regions and cultures of the world during the sixth century BCE.

Although each culture asked similar questions and offered valuable perspectives and insights toward understanding our world, it was the pre-Socratic Greek Atomists of the sixth and fifth centuries BCE that laid the foundation for scientific thought for much of the two millennia that followed.

The pre-Socratic Atomists framed important questions about the nature of reality and perception. Their attempts to answer many of their own questions resulted in the establishment of paradoxes, many of which continued to be heated sources of debate in venues as diverse as philosophy and psychology classrooms to high-energy-particle physics laboratories and deep-space astronomy observatories.

These paradoxes served two very valuable purposes to philosophers and thinkers and their students and followers. First, the existence of paradoxes indicated that the understanding of their world was incomplete or inaccurate. You might recall my earlier assertion that paradoxes don't exist in the universe; rather, they exist within us, the observers. Second, the paradoxes served as the starting point or portals by which, and through which, these philosophers performed their logical analyses of their world. Through nothing more than logic and common settings in their environment, these Atomists performed original thought experiments and developed a chillingly accurate understanding of the nature of matter that could only first be definitively confirmed using the technology of the twentieth century, when individual atoms were first imaged by the diffraction of x-rays and then by atomic force microscopy techniques. Though incomplete, the philosophy of the Greek philosophers and the scientific understandings of the nature of matter of the twentieth century are quite similar in many regards.

The Slippery Question of Fish Moving through Water

Allow me to create in your mind the image of a fish swimming effortlessly through water. For most people, such an image conjures up feelings of relaxation and serenity and the sounds of gurgling water, but not for the early Atomists! The concept of a fish swimming through water became the essence of a heated debate that arguably continues today, nearly 2,500 years later.

During the fifth century BCE, Leucippus became the first Greek philosopher to develop the theory of atomism. For whatever reason, the thoughts and teachings of Leucippus didn't make it directly into the historical record; so much of what we know about Leucippus and his theory of atomism came to us through his better known pupil, Democritus.

Upon reflection by the early philosophers, a fish swimming through water presented a paradox about their world involving the nature of matter. What resulted seems absolutely astounding to me. A seemingly mundane setting provided a laboratory sufficient to these early philosophers in which to examine and understand so much about the fundamental properties of matter.

The paradox can be described as follows: If both the fish and the water are comprised of matter, then the fish should not be able to move forward through the water unless and until a void forms in front of the fish into which it could then move. Yet, the fish apparently travels through the water without too much of a struggle. Without an empty space into which to move, the fish should remain motionless and frozen in its place.

To overcome this and similar paradoxes, Leucippus, Democritus, and other Atomists proposed that the matter comprising the fish and the water must be composed of very small, indivisible particles surrounded by empty space. As a consequence, they went on to propose that the natural world was comprised of two fundamental components.

These philosophers referred to the small, indivisible particles of matter as *atoms*; the empty space representing the absence of atoms was referred to as *void*. (The word "atom" comes from the Greek word *atomos*, which literally translates into English as "not able to be cut.") Today, we have identified particles that comprise the atom, so the contemporary use of the word "atom" is different than what the Atomists had intended. The known and scientifically accepted particles are represented by the Standard Model and are described later in the book.

Interestingly, Parmenides, a colleague of Democritus, completely rejected the notion that perception through our senses could be trusted to provide meaningful information. Instead, Parmenides believed that only logic and abstract reasoning could lead to truth and understanding. Incidentally, Parmenides was skeptical about the "existence" of the void. I will return to discussions about the void throughout this book, since the role of the void is crucial to a good understanding of our universe.

The deductions of the Atomists through logic alone borders on the incredible, in my mind. Twenty-five hundred years ago, human logic deduced the existence of indivisible particles of matter that could combine into different clusters to create the matter we observe through our senses! The fact that the Atomists accomplished this—with a laboratory consisting of nothing other than a fish, a pond, and their minds—only makes their accomplishment that much more dramatic.

The work of the Atomists, then, was notable for a few reasons. First, the Atomists made the distinction in their model between that which is matter and that which is *not* matter. Second, they proposed that matter is comprised of minute, indivisible particles that are capable of combining in different ways. Third, they advanced the concept of the void.

Interestingly and seemingly appropriately, the first and second achievements of the Atomists set the foundation for work that continues unabated today. Now, we probe the nature of matter through the use of advanced computational modeling methods and contemporary high-energy particle accelerators, including the Large Hadron Collider, which will likely remain the most powerful accelerator in the world for some time. We use these tools in an attempt to identify once and for all the most fundamental particles that are the building blocks of matter, the particles that the ancient Greek Atomists referred to as "atoms."

That Pesky Void

The third aspect of atomism, namely that a void surrounds the individual atoms or clusters of atoms, requires further clarification. Language is a very useful tool for communicating, but oftentimes words are either too limited or too ambiguous in their meaning to fully express a concept. There's a fundamental problem with words: Those that are available for us to use at any given moment to describe existing objects, phenomena, and concepts or past events are not always adequate to describe newly observed objects and phenomena or newly developed concepts.

Words generally best describe the attributes of an unchanging world or events that have already occurred. The problem for philosophers, scientists, engineers, and others in the business of developing new concepts, inventions, and products, is that it's exceedingly difficult to frame new concepts when the words do not yet exist to properly describe the concept or invention. At the risk of stating the obvious, nothing can be built until it is first conceived. New words must often-times be created to accommodate the emergence of new concepts and inventions. Researchers must never be shy of doing this, because it's necessary if new concepts are to be adequately understood and expressed.

Atomists had difficulty in verbalizing their concept of the void since the language of everyday human experience was not fully up to the task. In the minds of Atomists, the void represented the absence of matter, which itself took the form of atoms. They didn't mean, or want, to ascribe the property or state of existence to the void. To state it so completely violated their concept of the void. The void to them was not a state of being; it was a state of non-being. Sure, it was the opposite of matter in that it had no state of being. Yet, it's nearly impossible in English to refer to the void without using some conjugated form of "to be" or "to exist" in the statement, as in: "The void *is* the absence of matter."[9]

To refer to the void as a state of being was anathema to these philosophers. It would be more appropriate to describe their view on the void as, "The void is not."[10] If the concept of nothing is said to *exist*, then this concept can surely not refer to their concept of the void. The void is represented by utter nothingness, the likes of which may be implausible for human minds to fully embrace. The exercise to comprehend nothingness becomes another futile attempt by the mind to comprehend infinity, just in a different form. And, as you might have guessed by now, the concept of infinity exists nowhere else in the universe but in the mind.

Here's a bit of a brainteaser and another way to think about the Atomists' view of the void, given the limitations of the English language: It might be con-strued that the Atomists perhaps believed more in the *absence* of the void than in the *existence* of the void.

It's this third element of atomism that requires a bit more scrutiny. Hav-ing satisfied myself that I've adequately described the non-being aspect of the void, it would appear somewhat difficult for someone to focus their attentions and efforts on such a non-entity. From the perspective of the Atomists, the void is not something to study; rather, it's the quintessential *absence* of something to study. Although much work was performed over time on the nature of the void, its elusiveness has resulted to a large extent in the acceptance of ambiguous and incomplete understanding of it. The *matter* portion of our universe has been far easier to focus on and its *apparency* has made it far more attractive for researchers to pursue. After all, in the modern world, it's far easier to write a research grant on something that reviewers and funding sources accept and perceive as real than on that pesky, ambiguous void stuff.

Needless to say, we'll be revisiting the concept of the void later. It would be completely futile to understand our universe without coming to grips intimately with the void.

Chapter 11

The Universe from the

Perspective of a Photon

More than one scientist has considered making an attempt to comprehend what the universe might look like to a photon. Never mind that the photon doesn't possess a sensory apparatus with which to perform observations, perceive perceptions, and construct some sort of interpretation from which it assembles a model of the universe. Nonetheless, let's confer these attributes on a photon and see where it takes us.

Let's perform a short review of the characteristics assigned to the photon by science. The photon consists of an electromagnetic wave function with an indeterminate set of values that collapse down to determinate values upon measurement by an observer. As a quantum of electromagnetic energy, the photon is a packet of "light," although we understand by now that by use of the term "light," we don't mean that the wavelength of the photon is required to occupy the narrow portion of the electromagnetic portion we refer to as "visible light." As a photon of "light," the photon travels through free space, quite appropriately, at the speed of light that science accepts as possessing a value of about 300 million meters per second. Since the photon possesses a zero rest mass, this packet of energy can travel at the maximum limit allowed it in free space, again the speed of light.

We need to review what Einstein's relativity tells us about the unique clocks assigned to each observer and see how it applies to our anthropomorphized photon. Relativity tells us that the yardsticks we apply to space and time twist and bend while traveling at relativistic speeds—in other words, at speeds that approach the speed of light. Many of us have heard the interesting hypothetical story of an astronaut who departs Earth on a voyage at relativistic speeds and returns to Earth later to discover that he has aged far less than his peers and family members. The result of traveling at speeds far faster than the others he left on Earth resulted in his covering far greater distance during any increment of time than the people who remained on Earth. At those speeds, distances shortened according to an equation devised by Hendrik Lorentz during the late 1800s, which is referred to as the Lorentz transformation. The equation looks like:

$$\gamma = \frac{1}{\sqrt{1 - \frac{v^2}{c^2}}}$$

In this equation,

$$\gamma = extent\ of\ spatial\ or\ temporal\ contraction$$

$$v = velocity$$

$$c = speed\ of\ light$$

According to the Lorentz transformation, space warps so that distances in front of the astronaut shorten, while those behind him lengthen. Notice that I could have said that the astronaut *perceived* that space warped. From his perspective and the conditions that accompany it, this warpage of space is real to the extent that absolutely any means to which he might have access could be employed and analysis of any objective scientific measurements he might accumulate would match his perception that space really warped. Any non-relativistic spectator/observer would not "see" any such morphing of space around the astronaut's vehicle when the appropriate information finally reached the spectator's eyes or instruments at the speed of light. What different observers see is relative to their perspective and motion, hence the theory of *relativity*.

Let's return to the perspective of the astronaut travelling at relativistic speeds. Use of the Lorentz transformation informs us that space in the direction in which the astronaut is travelling has shrunken considerably. For the speed of light to remain constant to the astronaut, time must be passing more slowly, since the fraction

$$\frac{distance\ that\ light\ travels}{time\ over\ which\ light\ travels\ this\ distance}$$

must remain fixed at about 300 million meters per second. Since a fraction is nothing more than a ratio, if the numerator gets smaller, then the value of the denominator must also shrink, if the ratio is required to have a constant value of 300 million meters per second. The Lorentz transformation allows us to calculate the warpage of time or space directly based on the astronaut's speed.

Let's wrap this up so we can return to what a photon would see as it peered with awe at its universe.

As the speed of an object approaches the speed of light, spatial length and time contract around the object in such a way as to ensure that the speed of light is preserved at its fixed value of 300 million meters per second. So, then, what happens to time *at* the speed of light? Let's return to the Lorentz transformation to calculate this.

The Lorentz transformation for space or time is:

$$\gamma = \frac{1}{\sqrt{1 - \frac{v^2}{c^2}}}$$

When the velocity reaches the speed of light, c, the equation collapses to:

$$\gamma = \frac{1}{\sqrt{1 - \frac{c^2}{c^2}}}$$

$$\gamma = \frac{1}{\sqrt{1 - 1}}$$

$$\gamma = \frac{1}{\sqrt{0}}$$

$$\gamma = \frac{1}{0}$$

The value of γ then approaches infinity and the dimensionality of space and time collapse.

The result is clear and unambiguous. At the speed of light, space and time both become undefined and cease to exist. As startling as that may appear, let's now apply what we've learned to the perspective of the photon. The photon is travelling at the speed of light. At the speed of light, the length of space collapses to zero. At the speed of light, time collapses to zero. From the perspective of a photon, the universe consists of a single dimensionless point that it occupies and for which there can be no passage of time. From the perspective of a photon, there is no space and no time.

Let that sink in a while. Not only is it deeply troubling and deeply fascinating, it is also deeply consistent with many troubling aspects of the universe we've been discussing, including Feynman's Sum-Over-Paths model, which he proposed to explain the results of the double-slit experiment. If time and space are not fundamental exclusive properties of the universe, the facts that (1) to us it appears as though the photon took every conceivable path through the universe on its journey from our lamp through the slitted panel to the screen, and that (2) to a photon the universe appears as a dimensionless, timeless point might not be as preposterous as they seem either separately or jointly. In fact, there is a way to make them perfectly consistent. Just be patient, we're getting there.

Chapter 12

Getting Ready for New Thinking

Quite frankly, I've exhausted my inventory of delay tactics and have run out of ways to further delay the inevitable. So, I suppose it's time for me to begin introducing some original new ways to view the myriad phenomena we've so far discussed. Allow me to summarize some key paradoxes and phenomena before I proceed:

1. Recent experiments and observation reveal that the universe contains far greater mass-energy than we can directly observe.

2. Quantum mechanics indicates the natural, unobserved state of a quantum is a probability wave that possesses indeterminate characteristics prior to measurement by an observer.

3. Conversely, the act of performing an observation of a quantum and taking a measurement causes it to collapse to a determinate state with the particular characteristics we observe.

4. Relativity demonstrates that space-time is in the eye of the beholder and not a fundamental, independent property or set of properties belonging exclusively to the external universe.

5. Entanglement shows us that spatial separation is not what it appears to be.

6. The results of double-slit experiments indirectly indicate the presence of uncollapsed probability waves representing quanta that we cannot observe directly.

7. Calculations based on measurements of the amount of empty space in "solid" matter indicate that there is only an incremental difference between empty space and what we call "solid matter."

8. The best we can do as observers is to perceive perceptions, and these perceptions can be way off the mark regarding the "actual" state of things, like the universe, for instance.

9. All measurements of gravity indicate that space curves only in one direction and that's inward.

10. Science has yet to understand the nature and structure of the void.

There are, of course, numerous other items I could have included, but this small list will suffice for now. After all, I've attempted to diligently, deliberately, and faithfully reproduce within this book numerous scientific concepts, mostly in the form of reasonably simplified models of those concepts. I've also tried to convey an entertaining approach to questioning what we think we know in order to better understand the difference between the way we perceive things and the way they "actually are" from a "scientific" perspective based on limited and biased human sensory experiences.

Together, we've encountered and discussed a good number of scientific theories, phenomena, history, and perspectives. Together, we have come to understand to some useful extent the limitations of today's science in addressing key aspects of our universe. We have also come to understand the implausible and possibly flawed nature of certain current scientific interpretations—those that cannot account for numerous important features of our universe, for which ample, precise, scientifically based data have already been captured.

I've also shared with you my perception that the inexplicable paradoxes of our universe may arise from the prospect of incomplete scientific understanding combined with the likely prospect of flawed scientific interpretations that contaminate the current scientific understanding of our universe. I proposed that such a possibility arises from imperfect scientific methods performed by imperfectly objective scientists possessing biased views from which no person is immune.

I've introduced the concept that no object, theory, or mathematical model can be constructed before it is first conceived. As a result, science is largely limited to designing and building tools, which are useful to examine a universe that conforms to the prevailing notions of scientific culture. We've also discussed the propensity for human cultures to form rigid belief structures, which can be preserved far beyond their usefulness; science is no different in this regard. We've discussed the ways in which this propensity manifests itself, oftentimes, as a form of defensiveness to preserve the current culture, even when analysis of the data that the culture itself collected would appear to be in violent opposition to prevailing cultural beliefs.

Finally, we've confronted some of the fundamental limitations of the scientific method and its total inability to prove the non-existence of objects and phenomena, for which direct observation is currently impractical or implausible.

So, the hope of overcoming the hurdles to developing an improved understanding of our universe rests, in my opinion, on humanity's ability to propose bold, new ways of thinking that assemble the pieces provided to us in the form of unbiased scientific data into pictures that might look startlingly different from anything we have encountered or considered previously. Yet, the culture of science and its methods do not encourage the emergence of such radical thinking and theories (for which a deliberate and tangible connection to prior, accepted theories, interpretations, and conclusions already exists). Of course, this statement is also untrue to the extent that a willing individual may, at a time and place of his or her own choosing, introduce new thought frameworks in an unorthodox way, such as through writing a book, like the one you're reading right now.

Having said this, I now feel comfortable to share some new ways of thinking and thereby introduce the prospect of a new framework by which to consider, examine, and perceive the nature of our universe. Perhaps these concepts and their associated thought-process frameworks will contribute to breaking the deadlock of the last few decades.

For much of the remainder of the book, I will first propose new hypotheses to address the inadequate explanations currently offered by the prevailing scientific orthodoxy, and I will then follow each hypothesis with a discussion of its consequences in an attempt to reduce the hypothesis to an unacceptable level of absurdity. Some of these hypotheses may at first seem preposterous, yet we will discuss the ramifications of each hypothesis until we can either conclude the discussion with absurdity or not. I will choose to accept as plausible those hypotheses that survive.

Throughout this process, please remember that I am only attempting to propose, or produce through iterative discussion, plausible concepts to address and eliminate the paradoxes that emerge from the current state of scientific understanding. The reader should not interpret these plausible explanations as scientific truth. Scientific truth can only be established through a thorough, cautious, and deliberate process of experimentation, observation, and interpretation followed by a thorough process of peer review.

With these disclaimers now behind me, let's get going.

Section II

<><><><><><><><><><><><><><><><><><><>

Correlation, Integration, and Formulation: The Beginning of New Understanding

Chapter 13

The Speed of Information Revisited

Let's pull out some of the interesting observations and conclusions from our previous discussions and add into the mix some of the relevant thoughts and concepts introduced by key scientists and philosophers. In this way, perhaps we can combine our perspectives with those of others to construct a singular framework in which each of these key insights may coexist harmoniously.

The results of our experiments with entangled particles throw into question our most basic notions of time and space. Entangled particles appear to have the capacity to communicate with their partner particles at a speed far greater than that at which electromagnetic waves can propagate through empty space. In fact, the possibility that the communication occurs instantaneously cannot be ruled out by current scientific measurements. This observation has become quite a contentious point within the science community because it would appear to violate important aspects of relativity that limit the speed at which information can be communicated.

In fact, a substantial number of scientists have rationalized that light speed remains the speed limit for information communication. The scientists who believe this have constructed elaborate schemes and justifications to support their continuing contention that information simply cannot and will not be communicated faster than the speed of light.

With all due respect to these scientists, this is not the way I choose to see it. In fact, I believe their defense is built using science and math in the mode I described earlier, in which facts are assembled to support a biased and flawed view and to support a conclusion chosen beforehand as the "right" answer.

Allow me to explain how I view the situation. I'll set up a small thought experiment to assist me with this. I'll use my lamp and my crystal to produce a pair of entangled sibling quanta. I make absolutely no attempt to measure my quanta, so they remain in the form of indeterminate probability waves. Please recall our discussion of the Heisenberg Uncertainty Principle. If, indeed, we attempted a measurement of one of these quanta, the probability wave would collapse to a single state and the resulting determinate state would be captured by our measurement and recorded. If we determine the precise position of the resulting particle, then we have little clue about its momentum and, consequently, we will have little idea about its whereabouts and where we might next find it.

Okay, so we take one unmeasured quantum and place it in a special jar that's sufficient in size to ensure that the "photon" is inside and we set it aside for the moment, where our quantum will remain safe and sound. We let the sibling quantum escape into the universe. We had already left the doors and windows of our lab open and asked all of our colleagues to look the other way and shut off their detectors to ensure that the "particle" remained unobserved. Perhaps, we kick back for a few minutes and enjoy a cup of organic, fair-trade java.

When we're good and ready, we retrieve our jar and perform a measurement of the captive quanta, collapsing it to a measureable determinate state, and we observe the photon to possess a spin value of "up." We know, for sure—positively no bones about it—that at the precise moment that we measured our captive quanta, thereby forcing it into a measureable determinate state, that the sibling particle collapsed down to a measureable, determinate state possessing a "down" spin value, no matter where that sibling photon might be. I now possess information about the state of the runaway sibling particle and perhaps no time was involved in that information reaching me.

Now, it's true I don't know as much about that particle as I would really like to know, such as its whereabouts or where it's headed, but I know its state with complete certainty. I acquired knowledge of its state "instantly." There was no delay, as I would expect of information that reaches me at the now seemingly sluggish speed of light. In essence, I performed my measurement of the distant, runaway quantum by observing the captive quantum at hand.

Please note that an expanded but pretty equivalent version of the Heisenberg Uncertainty Principle remains very much at play. The expanded version of this principle could be described along these lines:

"The more precisely I know the spin state of an entangled particle, the less I will know about its other features, such as position and momentum, and those of its entangled sibling." There may be many variants of this expanded Uncertainty Principle that others might devise, but none of them would deviate in any significant way from the intent of the original.

So, if you buy into my arguments, it would appear that there's some question about the speed limit imposed by our universe on the propagation of information. All of the information that we create and communicate in our increasingly technologically complex human culture is subject to the speed of light limit—of that we can be certain. I personally can't think of a single exception.

Nonetheless, our thought experiment provides us with a simple, clear, concise message. It is possible for information to break the limit imposed by the speed of light. But, we need to make one important, additional distinction: Information is limited to the speed of light if the propagation of that information relies on the electromagnetic force.

Let's review our four fundamental forces from the perspective of communicating information:

1. The strong force is impractical to use as the basis for a communications scheme unless we resided in the nucleus of an atom due to the exceedingly small distance over which it operates.

2. A very similar thing can be said of the weak force, so it can't be used in a practical way for communication.

3. The electromagnetic force can be used to transmit information up to the speed of light but no faster. Electromagnetism is a very practical basis for communication, but the fact that its strength diminishes rapidly with the square of the distance makes it really frustrating to use as a long-distance communications scheme. Over very large cosmic distances, the length of time for communicating information becomes unbearable, and the signal strength diminishes to a very, very low level.

4. Gravity represents another prospect to serve as the basis for a communications scheme. The problems with using gravity, however, are probably abundantly obvious. First, no one has ever observed a gravitational wave and so we can't even be sure that such things "exist." Scientists suspect that gravitational waves, carried by the "mathematically devised but never observed" carrier particles of gravity called "gravitons" are also limited by the speed of light, so gravity would leave us with another slow-pokey communications scheme for long distances.

5. Finally, add into the mix that gravity is outrageously weak and that its strength also drops off with the square of the distance, and what results is probably a really poor, ineffective communications scheme. The only possible thing going for gravity is that it can't be blocked

from reaching us, no matter how puny its signal. Even though the concept and mathematical basis for gravitational waves have been thoroughly investigated, they have yet to be observed. The measurement apparatus could also be quite large to provide a desirable sensitivity level, like something on the order of the diameter of our solar system, rendering the discussion somewhat moot for the time being. The concept and mathematics of gravitational waves have been successfully used to understand the decay of the orbit of one celestial body around another and this is an area of science where further progress is likely needed.

So, the bottom line regarding the four fundamental forces is that, in and of themselves, there really aren't any particularly good choices to serve as timely and practical long-distance, cosmic, communications schemes.

On the other hand, however, we have entanglement. Could it serve in some way as the basis of our long-distance cosmic communicator? Could entanglement be the basis of a *Star Trek*-ian subspace communications scheme?

(There it is. It's out in the open. I too, grew up on a steady diet of *Star Trek*. By the way, *Star Trek* provides the most striking examples yet of some concepts I introduced earlier, namely that no concept or device can be built without it first being conceived, since its entry into the universe begins with the emergence of a concept that exists nowhere else in the universe but in the mind. The gadgets used in the Star Trek series have motivated the development of so many of the devices we take for granted today, such as flat panel displays, iPads, iPods, iPhones, and magnetic resonance-based medical imaging devices representing some of the clearest examples.)

To answer our question about whether entanglement could be useful for communication, we need to recall the discussion we had a few minutes earlier about not knowing where in the entire universe the companion particle of a generated entangled pair might be. We'd certainly have a losing proposition if our scheme involved as its sole basis first producing entangled quanta pairs locally in the lab and then letting one member of each pair drift aimlessly through the universe to reach a distant receiver. First, the photons could only travel up to the speed of light and, therefore, it would be as slow-pokey as just using electromagnetism in the first place. Then, we wouldn't know where our partner particles were, so we wouldn't be able to pick and choose with what portion of the universe we would communicate. Okay, so that idea was dead on arrival. Well, maybe, there's something we're not considering.

Let's talk about the nature of entanglement. Is entanglement simply a laboratory curiosity? Science has made numerous attempts to answer this question, and the answer appears to be no, it isn't. In fact, scientific experiments and the data they've produced are quite clear on this question. Entanglement occurs everywhere scientists look for it.

Although Einstein did not believe in the existence of entanglement and felt that quantum mechanics could not permit the interaction between two spatially non-local objects or phenomena, the more science looks for entanglement, the more entanglement it finds. Using crystals to produce entangled pairs in a laboratory setting from a parent photon is only one way to produce entangled particles. It turns out that the universe just spews out entanglement at a frantic rate. All that nature requires is that two like particles get spatially close enough for them to "touch" and interact with each other. After that, these particles possess an entangled relationship until one or both particles engage and "touch" other particles, at which point the original entangled relationship between a particle and its partner is replaced by the new entangled relationship it forms with its new partner. Since random collisions between particles is pretty much the usual state of affairs in most of the universe, entanglement can aptly be described as universal. This is an important concept; entanglement appears to be everywhere.

Does this mean that each and every particle is entangled with at least one other particle somewhere in the universe at any particular moment in time? No, I wouldn't necessarily think so, and this question involves a premise that is impractical for science to address. Science has no way to observe and characterize the presence or absence of entangled relationships among each and every particle in the universe at one particular moment in time.

Okay, so is there a different question that science can answer? Remember that science is limited in what it can precisely measure because of the Heisenberg Uncertainty Principle. We're limited in the knowledge we can accumulate, since we don't know where to find one or all of a particular particle's entanglement partners. In fact, science would be hard-pressed to take what we already know about entanglement much further in the near future. Hopefully, clever scientists and their tools will provide us with more insight into entanglement over time.

So let's recap what we know about entanglement. It is a natural and common component of our universe. While not particularly well understood, its prevalence in our universe is well-established. Entanglement indicates to us that our notions of time and space may be deeply flawed. A limited amount of information (like quantum spin) can be transmitted quickly, perhaps instantaneously, between partners engaged in an entangled relationship.

This, in turn, causes me to question the foundation for our belief that information cannot travel faster than the speed of light according to Einstein's Theory of Relativity. That is, unless the particles are not actually spatially separated, which would explain the part about breaking the limit set by the speed of light, though it leaves us at a loss to explain how two particles apparently separated by space are, in fact, not separated at all. We also know that entangled relationships exist at least among many but perhaps not among all particles, quanta, photons, and "what-not" out there in the universe.

Chapter 14

Existence versus Observation and the

Two Types of Observation

N ow, let's transform some of the wording above to detect the presence of objects or phenomena through the perception of signals sent using the electromagnetic force or "light." At the moment we take our measurements and perform our observations, the quanta of energy that we, in fact, observe are rendered visible (detectable, measureable). Through the actions of our observations, the quanta we are able to observe collapse from an indeterminate probability wave to a photon possessing determinate characteristics. We refer to the phenomena, or object, and the electromagnetic waves from which we've learned about the phenomena or object, as "luminous."

Let's take this concept and analyze it for another moment. What does this really mean? Could we have learned about the phenomenon or object another way? As we previously discussed, of the four fundamental forces, only one is suitable and practical for us to use as the basis for a meaningful communications scheme. We could put this more forcefully, however, and simply state that, for all practical purposes, the only means by which we perform "direct" observations of anything is exclusively via the electromagnetic force, and this is the *only* force upon which our senses operate and rely.

Let's get even more forceful and say that the only means by which we can perform "direct" observations requires the electromagnetic force as an intermediary between a phenomenon or object and our senses. I think I'm now pleased with this wording. Therefore, the one thing common to all of our observations is the electromagnetic force acting as an intermediary between "it" (whatever "it" is) and us. Thus, electromagnetic force is the sole basis for each of our perceptions through each of our senses: taste, sight, smell, sound, and touch.

We are also aware of the results of the single-photon, double-slit experiment talking to us in a clear, unambiguous voice that there are other uncollapsed, indeterminate probability waves that remain otherwise undetectable to us through direct observation, as evidenced by the presence of continuing interference patterns.

If you follow along with each of these arguments and agree as to their reasonableness, separately and collectively, then we are left with an interesting dilemma. It was Einstein himself who said that if you've eliminated all of the other possibilities, then no matter how implausible the remaining possibility might be, it must be right.

So, what is our sole remaining option? I submit for your consideration that not every photon we attempt to measure can be detected and subsequently measured. As scientists attempt to perform observations, there are those quanta that collapse down to detectable, determinate states and there are those that don't. If they're close enough to us, there isn't a single effective, meaningful scientific instrument that can tell us that those quanta are in fact there.

I suggest that right now is a really good time to take a break and let this statement sink in. Get a cup of coffee, go for a walk, enjoy some music, and check your e-mail, perhaps, but give yourself enough time for your mind to wrap itself around this concept and allow this interpretation to become part of your internal universe.

...

Now that we're back, let's talk this out some more.

Upon what basis might it be possible for a quantum to be rendered undetectable? The short answer might involve entanglement.

Do you recall the thought experiment in which we produced an entangled quanta pair, placed one in a jar, and let the other partner particle fly free into the external universe? Let's alter the thought experiment. Let's place one quantum in our internal universe and its entangled partner particle into the external universe.

There's a barrier that each of us defines as the dividing line between the "self" (internal universe) and the rest of the universe or the world around us (external universe). It might be helpful if we reproduce part of the glossary here for convenience, so you can review some of the definitions I previously proposed. We're going to use these definitions while performing our next actions. As you might also recall, the boundary between these universes is not a sharp, well-defined boundary according to quantum mechanics. Rather, the boundary is squishy and vague on a quantum mechanical scale.

Here's a review of the terms so far in our glossary:

◊ **Consciousness** – the act of synthesizing information about the internal self with information about the external universe.

◊ **Life** – (a) the state of matter that meets all three of the following criteria: (1) possessing self-awareness of internal states, (2) capable of perception, and (3) incapable of existing in an indeterminate state with an uncollapsed probability field since the state of matter comprising the living entity is under continuous self-observation; (b) the state arising from a self-observable arrangement of matter resulting from self-awareness of internal states.

◊ **Non-life** – inanimate objects and/or phenomena that meet all three of the following criteria: (1) not possessing self-awareness of internal states, (2) incapable of perception, and (3) capable of existing in an indeterminate state with an uncollapsed probability field when not observed.

◊ **Observation** – the synthesis of information by a living entity that connects self-awareness with specific information it perceives about the internal self or the external universe.

◊ **Observer** – a living entity capable of perception relevant to a specific observation or set of observations; for example, an observation involving electromagnetic energy requires light perception in the correct portion of the spectrum by the observer.

◊ **Perception** – the collection of information by the self about the internal self (internal states) or about the external universe (external states) through the capabilities of its unique sensory apparatus.

◊ **Self** – a living entity capable of conceptualizing a boundary distinguishing that which it considers internal ("itself") from that which it considers external ("the world" around it).

◊ **Self-Awareness** – the sense of self unique to each self.

As I said, let's place one particle of the entangled pair on either side of this boundary: one within our internal universe and one within our external universe. In this configuration, I propose that our mind continues to serve as it always does in the capacity of "bridge" between the internal and external universes. Our mind serves as the conduit between what we can sense about the universe (perceptions) and the interpretations we construct and the understandings we achieve (perception of perceptions). The understanding we achieve allows each of us to construct a model of the universe in which to accumulate and integrate our experiences, and to place them into context with one another. It is our mind, then, that rightfully or wrongfully defines our sense of relationship with the external universe and our sense of reality.

I propose that the basis of the link between "observer" and "phenomena" is defined at its most fundamental level by quanta that collapse to determinate states, which can then themselves be measured. And whether or not a particular quantum can and will collapse from its indeterminate, probability-wave, quantum-mechanical description down to a determinate, observable state may be defined by the presence or absence of an entangled relationship between the observer and the photon upon which the transfer of information vitally depends. The presence of an entangled relationship between the observer and a particular relevant quanta of electromagnetic energy may result in an intimate relationship that consequently collapses down to a determinate set of states, ending in the successful flow of information and the start of the perception process. The absence of such an entangled relationship between an observer and any unrelated indeterminate quanta might result in no collapse of state, no detection, no flow of information, and complete and utter obliviousness to particular objects or phenomena surrounding us in the external quantum universe.

Before we go any further, I really want to have a little chat with you so I can get something off my chest. Please understand that I know that the ideas I proposed a few moments ago are peculiar; in fact, they're really peculiar. I realize that you may be resistant to these new ideas, and you might find them uncomfortable. You might even react more strongly to them. You might find them incredulous. They might even evoke downright hostility in you. Frankly, I share your incredulity and skepticism; it's a bit unnerving for me to share such radical concepts. But, I can also share some things for which I'm positively confident. These concepts are no stranger than those with which we've been living and accepting for decades now, and these new ideas offer a way out of the current scientific paradoxes with which many of us are sick and tired of living. In fact, these concepts may even be simpler and less peculiar than the ones we've accepted. Remember that the concepts we've previously accepted must be flawed and/or incomplete if the picture they form of the universe is unequivocally wrong based on the paradoxes that are thrust upon us.

I'm proposing a new framework that many will consider bold and possibly flat-out wrong. History is fraught with examples in which bold, new thinking was met by resistance of the culture, more often with open hostility than not. In many ways, the emergence of this resistance and hostility is the beginning of a thinking revolution. Some aspects of the framework I propose may be greeted with acceptance and others with skepticism and open criticism.

Nonetheless, the goal of my book and the concepts I propose is to *shake us out of our stupor and toward a new state of thinking, with the prospect that the next iterations and generations of thinking will propel our awareness and understanding of the universe, and that a picture of our world might eventually emerge that more accurately reflects the truths and realities of a universe that is loathe to exposing its internal workings, especially given the narrowness of our senses.*

We will continue our voyage and see how these new concepts join with others I have yet to present in such a way as to eliminate the paradoxes we so abhor.

Let's now focus on the ramifications of undetectable objects and phenomena in our general vicinity. As we've encountered so far, the strong and weak forces are each very nice forces, but they're completely useless to us in our present context of trying to detect unobservable stuff around us. If our hypothesis of unobservable, persistently indeterminate, electromagnetic quanta is correct, then indeterminate packets of energy exist and are loitering around us, and the possibility of using electromagnetism is right out, since it provides a deceptive scheme for observations of this type. Gravity, unfortunately, is also useless to us, since we have no reliable way to use it for local, small signal measurements. In fact, we are once again in a cosmic pickle, but only from the perspective of making a local measurement.

Recall that we can't use science to prove the non-existence of an object or phenomenon. Of course, science also cannot be used effectively to measure the existence of unobservable phenomena or objects when the spectral sensitivity (capabilities) of the sensors are incompatible with the phenomena or object we wish to observe. In other words, if our ears don't operate within the same portion of the acoustic spectrum in which a falling tree produces its vibrations, we don't hear the sounds. We, therefore, can rightly say in such a case from our perspective that the falling tree produced no sound even when an observer is standing right off to the side of the falling tree.

So we need a way to measure unobservable objects or phenomena. The strong, weak, and electromagnetic forces remain off the table for the reasons we already discussed. We also can't use gravity to perform a direct observation; however, we're very clever, so we'll devise a new scheme. In this scheme, we'll combine gravity and electromagnetism together to permit us to perform indirect observations involving distant matter and space.

The scheme looks like this. We'll use electromagnetism to observe distant luminous matter that's spinning around other luminous matter. By making these observations, we'll measure a few things: the amount of luminous matter, the rate at which these clumps of matter spin around each other, and the distance separating the centers of gravity for each clump of stuff. We'll use entire galaxies as the basis for our observations.

This scheme seems okay, even though it depends critically on our ability to observe photons. Actually, this is fine, since we can depend on measurements based on photons so long as we can detect them. And, in our experiment, we are performing observations only on the photons we can detect, of course. So, for the time being, we can forget entirely about the ones that remain unobservable.

Now we go off and take our measurements using all the scientific and technology tools available to us, including the most recently developed and deployed space-based observatories, such as the Infrared Astronomy Satellite (IRAS), the Cosmic Background Explorer (COBE), the Infrared Space Observatory (ISO), the Wilkinson Microwave Anisotropy Probe [WMAP, also known as the Microwave Anisotropy Probe (MAP) and Explorer 80], the Spitzer Space Telescope, the Herschel Space Observatory, and the Planck Spacecraft.

Once we've collected a mountain of data, it's now time to perform our analysis. Again, the measurements we've taken enable us to determine the amount of luminous matter, the distribution of the luminous matter in space, and the nature of the rotation of the two clumps of luminous matter with respect to one another.

Science has a toolbox brimming with useful mathematical tools in the form of equations and models. The equations and models govern the motion of one clump of mass around another. These equations and models are well-confirmed through numerous prior scientific observations, and the results obtained have been compared and correlated to results obtained through other equations and observational methods. There's a certain level of comfort we may permit ourselves to feel as we employ these equations. The equations allow us to calculate the amount of matter and the corresponding values of mass that each of the two clumps of matter possess.

When scientists take the data that they've collected through their instruments, perform their calculations, and then analyze the results, something very peculiar emerges, indeed. The results indicate that the clumps of matter—in this case galaxies—each contain roughly six times more mass than the direct observation of the luminous matter contained in each galaxy would otherwise appear to possess. These results only became possible during the last decade or so, with instruments possessing vastly superior technological capabilities than their predecessors of the late twentieth century, only a few years earlier.

Allow me, once again, to reword the results. The latest scientific results indicate that for each part of ordinary, luminous, observable matter that we detect in these distant cosmic clumps, there are five more parts of matter that are unobservable to our instruments through direct electromagnetic observation schemes. These additional five parts of matter can only be detected indirectly through their gravitational interaction with one another. Since we are unable to observe these extra five parts of non-luminous matter, scientists have termed the mysterious phenomenon "dark matter."

Recall that Einstein established through relativity theory that the basis of gravity is the inward curvature of space resulting from the presence of mass. Also, recall our discussion about gravity and space curvature—namely, that there is no scientific evidence to indicate that any sort of matter is capable of curving space in any direction other than inward. So, even though scientists have been utterly baffled by the existence of non-luminous matter, there is ample evidence that the phenomena involve matter combined with the traditional curvature of space, which occurs around the ordinary luminous matter that we are quite accustomed to observing. So, although this mysterious dark matter is non-luminous and not directly observable, it is also evident that the gravity it creates is completely old-fashioned, ordinary, and oh so (yawn) boring.

Chapter 15

Here versus There

Time for another recap. We can only observe ordinary matter directly through electromagnetism through the collapse of probability waves associated with quanta that collapse down to observable, determinate states that we can observe and measure. I hypothesized the prospect that undetectable, indeterminate probability waves associated with certain quanta do not collapse to an observable state and that this hypothesis is consistent with the results of countless variants of countless double-slit experiments.

If my hypothesis is correct, the phenomena are undetectable locally by any means currently available, other than the double-slit experiment. Furthermore, if my hypothesis is correct, the phenomena are detectable at a distance through indirect schemes involving mass and gravity. Recent scientific experiments indicate unambiguously that we observe non-luminous, directly unobservable, "dark" matter, and that there's a boatload of it. In fact, our observations enable us to calculate that in these distant regions of space there is five times more dark matter than ordinary, luminous, directly observable, "familiar" matter.

Of course, there's no scientifically based justification for "that" region of space to be any different from "this" region of space, so there's no basis to decide that one region of space is any more or less special than the other. Therefore, I would suggest it's possible that we only observe as few as one quantum out of every six or so quanta in our local vicinity. Perhaps entanglement is the basis for whether or not a particular quantum is observed, or perhaps it isn't. However, something is; that's for sure.

If indeed we observe only some portion of the quanta in our universe that choose to reveal themselves to us—and it certainly seems like that's the case—our resulting perceptions of the universe with respect to mass and matter may be very well off the mark. That might very well be an understatement when all is said and done. A universe comprised of far more matter than we had previously thought existed could complicate the scientific understanding of the universe and wreak havoc on the calculations that result from scientific mathematical tools. In other cases, the numbers may be correct, but the interpretations offered through science may require some major adjustments.

If our perception of matter is so far off, then it's little wonder why science has had such an extraordinarily difficult time of understanding and calculating the universal gravitational constant, as one prime example. Perhaps only when science further refines its ability to perform observations will it be able to improve its incomplete understanding of our universe to the point that its interpretations provide a more accurate representation of the subtle workings of reality in which paradoxes simply don't appear.

Herein lies a fundamental cultural issue. Through our many discussions, we've learned that science can't operate outside of certain, key boundary conditions within which it works quite well. External to these boundaries we learn that the scientific method can't be applied; the very possibility of using our beloved scientific methods, and the stunning instruments that our science has produced, is precluded. Sure, science can stubbornly go ahead and be implemented, but the results that science will produce under these circumstances become unreliable, yielding potentially false interpretations that might taint our understanding and contaminate the scientific database upon which science increasingly relies. Yet the results of this false science carry the full seal of scientific approval, with the result being a culture willing to adamantly and vehemently defend its assertions, even when being stared in the face by the paradoxes it produces.

So, through some combination of human civilization asking science to perform the task of explaining the universe to us and hard science's decree that it alone should and can perform this sacred task lies a dilemma for each of us to consider. What is within the realm of science and what is not? Society and science each demand that science produce a meaningful interpretation of our universe. Then we expect/demand that science produce answers even in the domains in which its very use is precluded.

Science needs to be more forthcoming in its limitations with respect to the dialogue it craves with the lay community. Certain aspects of science's rigidity and its crafted image of omniscience that it so adores need to be tempered to allow a freer exchange of ideas and speculations that supplement the positive aspects of peer review and open constructive dialogues. After all, science may not be able to provide all of the answers today, but it sure can ask all of the questions and be more receptive to the variety of thought processes and perspectives that emerge. Has our science lost sight of its own limitations in its mad, competitive rush forward?

A common cry in American society right now is that we need fewer politicians and more statesmen. A very similar thing could be said of our science community—namely, that perhaps we need fewer scientists and more philosophers. At the very least, must science preclude philosophy or view science's foundational underpinnings of philosophy as disdainfully metaphysical?

This seems like a really good time to lighten up and for me to propose our next glossary entry.

◊ **Matter** – determinate states in space potentially observable (measureable) locally by electromagnetism possibly in combination with the weak or strong forces, or potentially observable at a distance directly by electromagnetism in combination with gravity.

In fact, I'm going to let my hair down and introduce our tenth term, since I've already begun my introduction to the concept of space-time that I'll use throughout the remainder of the book.

◊ **Space-Time** – the temporal-spatial framework unique to each observer based on the inertial reference state of the observer.

If you read Dr. Robert Lanza's *Biocentrism*, you might notice some resemblance to the concept of space-time he introduced. Although others may have proposed similar concepts of space and time earlier, when I read Dr. Lanza's interpretation of space-time being the property of the observer, it struck me like a ton of bricks and set me firmly on the path of this book project.

We now have a glossary comprised of ten terms. Only eight more terms to go before it is complete.

In fact, I'll even tell you right now that remaining terms in the glossary include "void" and "mass" and that the universe seems to be whispering to us that their definitions might be far different than the ones to which we are accustomed.

I want to say one more thing before we go on to our next discussion. I've mentioned some concepts about the individual that in many ways require you to take very big leaps of faith, and then I forged ahead to maintain the momentum I was attempting to build. My intent was not just to drop them on you and run away. So I'd like to talk a little more about them.

Chapter 16

Symmetry, Physics, and Metaphysics

I discussed the prospect that the essence of living things that sets them apart from non-living things is their ability to perform self-observation. In essence, I mean by this that the mind establishes and maintains a set of determinate states within our "self," or our internal universe, in a scheme that thoroughly mimics the process in which the mind establishes and maintains determinate states in our external universe, a firm foundation for which is provided by quantum mechanics. Although some might consider that there's a metaphysical element to this, I maintain that this is absolutely not the case. Instead I'm proposing that the very same processes that quantum mechanics assigns to "out there" apply equally to "in here," using an argument often made by science based on the concept of symmetry.

Science has determined that nature positively adores symmetry and that science adores the proposition that nature adores symmetry. Symmetry often provides an effective means by which science creates simplified models of complex concepts through the use of analogies. The use of these analogies doesn't really differ in any substantive way from the literary use of the concept, for which the same term is used.

Most of us became acutely aware of the use and practice of analogies when we prepared for and took the Scholastic Aptitude Test (SAT). Yet analogies are used throughout our culture in order to convey such things as function, significance, and essence in a comparative way that allows our minds to build bridges between one concept and another. Scientists use the concepts of symmetry and analogy to construct a simple model of the fundamental particles and carriers of force, called the Standard Model. (See Figure 8.) In some cases, the use of these methods was the sole basis in predicting some of the entries in the Standard Model, and only through subsequent scientific research was the existence of these particles in our universe confirmed. In fact, at least some of these hypothesized particles remain to be observed and it is through the Large Hadron Collider that scientists hold out hope of their confirmation.

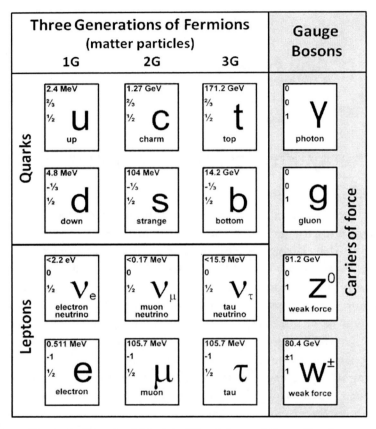

Figure 8. Standard Model of Particles and Force Carriers.

For my own arguments, I feel that I've made good use of symmetry and analogy. In each case, I feel that I've provided sound justification using methods embraced by the culture of science. Yet, it feels almost palpable even to me that some of these claims border on, or even cross over into, the realm of metaphysics. I discovered that I needed to review the steps of my logic over and over again before I achieved a level of comfort that satisfied me. Upon reflection, I was amazed at the extent to which reasoning that seems sound to me at first filled me with a sense of discomfort that I had wandered into the scientifically forbidden land of metaphysics. Only afterward did I realize the extent to which my experiences in the worlds of science and technology had walled off entire realms of thought in a synthetic, meaningless way. The experience has been extremely illuminating and personally gratifying but, at the same time, terribly difficult and at times downright painful.

Anyone, I believe, would be hard-pressed to deny the validity of the analogies I proposed. It seems to me to be quite reasonable that the properties and/or makeup of space over "there" should not be fundamentally different than those of space over "here." In fact, everything we feel we know tells us that the physics of "there" and the physics of "here" are the same. Even without any proof one way or the other, science must decide that the physics of "here" and "there" are identical, or else science would become completely paralyzed unless and until it could simultaneously observe all phenomena everywhere all at once to ensure perfect consistency and harmony within each of the nooks and crannies of all the prospects of the complete spectrum we call our universe. This situation, needless to say, would put a crimp in the methods and practice of science and prevent anything from ever getting done, even more so than today's institutional science bureaucracy.

Given the impracticality of this unspoken concern, science instead takes another leap of faith so it can then be practiced. Scientists have no choice but to *believe* that the same physics holds true at each moment and applies equally well in all places. This is a second element of science that reduces its distinctions from those of religion. The prospect that physics is inconsistent throughout the time and space of our universe is the "Voldemort" of science. And, like Voldemort, the possibility of its existence is highly unlikely and inconsistent with all that humanity has observed thus far.

So, you and I have little choice, really, but to accept that the physical world in distant regions of space is at least substantially similar to the regions of space closer to home, if not completely identical. I think most of us can feel comfortable with that. If "here" and "there" are the same, where is the boundary that separates the two? The very concept of a boundary begins to fade away as a useless, irrelevant accessory. The concept of "here" can only be distinguished from the concept of "there" because, put most simply, "there" doesn't seem to be "here." The only boundary that exists whatsoever is based only on the perspective of the individual self. This perspective exists nowhere else in the universe but within the mind.

Let's proceed with another analogy. By the very same token, since either the physics of "there" is the same as the physics of "here," or we have out of necessity chosen to accept it as the same, then the physics of the external universe cannot be distinguished, or permitted to be distinguished, from the physics of the internal universe.

This is where our cultural comfort zones begin to feel a little too warm for some of us. Ironically, it's the cultures of science and religion that get particularly rankled about now. This is yet another similar trait shared by both these cultures. I would really like to believe that the members of one of these cultures don't feel offended in any way by my comparison to the members of the other, but I consider the likelihood of that particularly slim. In the end, as well as at the start, we are all members of humanity and we share a bunch of common traits. Paramount among these is our desire to understand our universe and our place within it given the limitations of our senses and thoughts.

Let's try to deal with this sticky wicket now since it may become difficult for the reader to proceed to assertions, concepts, and content that appear later in the book until some basic thought processing is performed and some closure achieved.

All I'm saying is that the same physic applies on either side of the boundary defined by the self. We all seem to define such a boundary. This sense of boundary represents an important and intimate component of our unique, individual identity. Yet we've learned that the boundary separating "self" from the "external universe" is indistinct. There is no sharp, singular barrier separating the electromagnetic fields of the electron shells of atoms on the surface of our skin from those produced by the electron shells of atoms in the external universe. Indeed, the manifestation of the fundamental forces at work within our bodies intermingles with the manifestation of the very same forces at work throughout the universe. If this were not the case, we could not count upon other fundamental, well-understood, and observed physical and chemical processes and phenomena that are required to sustain our very lives and make it possible for us to even ask whether or not we exist.

Upon continued reflection of our human condition, there is little choice but to dismantle the wall that we falsely perceive as separating our "self" from the external world and "others" in precisely the same way that we have little choice but to eliminate the wall we have falsely constructed to separate "here" from "there" based on the faulty interpretations arising from our limited perception of perceptions. When we are ready to accept these concepts, we are far better prepared to meaningfully continue our quest to understand our incredible world and the place of each self within it.

Of equal importance, please note that I make no mention of a supreme being or deity in any of this. What my personal beliefs might be and whether or not a god of any sort or flavor exists is not relevant to the particular quest captured in this book. It is left to each individual to ponder the existence, absence, or nature of his or her god and the relationship he or she chooses to have with it. Even though science and religion are notorious for stepping on each other's toes, this book focuses on science and philosophy and will, to the extent possible or practicable, avoid treading on religious turf. At least, that's what I'm trying to do. I promise!

Chapter 17

Existence versus Observation

Let's take a moment to consolidate our perspective before we begin again to expand it further.

We've learned that the sum total of the mass-energy in our universe only ten years ago is a fraction of what we consider it to be today. By today's metric, the sum total of mass-energy we detected ten years ago represents only 4 percent of what science considers it to be now. The amount of ordinary, luminous mass-energy was based on direct electromagnetic observation. But, through newer scientific measurement capabilities, which combine electromagnetic-based observations with mass- and gravity-based assumptions, we now observe the summed total of mass-energy of our universe to be twenty-five times greater. These new studies are based on observing the movement of galaxies and other very large and massive distant celestial bodies around another, and combine these precise measurements with the use of well-understood equations to determine and interpret an accurate estimate of the mass of these hefty clumps of matter.

As a consequence, our science indicates to us that there is five times more dark matter than ordinary luminous matter. Our best science also indicates to us that there is more than four times as much dark energy as the total amount of luminous and dark matter combined. (Try not to think too much about dark energy yet; I will cover this topic in detail later.)

Now, let's put this into perspective with one of the key interpretations we discussed earlier, namely that the volume of a carbon atom is nearly entirely empty space. Let's now assign a similar value for each type of atom and particle and further assume that the same holds true of dark matter, especially if we accept my earlier proposition that dark matter is simply regular matter with which we possess no electromagnetically based observational relationship.

Our quick analysis leads us to the epiphany that for each and every morsel of matter, whether luminous or dark, there exist over 500 trillion morsels of empty space. The blatant truth is that empty space is far and away the biggest feature of our universe. Empty space positively eclipses any other feature of our universe. The big problem endured by empty space is that it's terribly boring to look at. Perhaps it's more accurate to say that it's terribly boring to *try* to look at it, since it possesses only a single attribute, which is that it is unobservable.

Undoubtedly, there are those who would take exception to my description, but they're just being sticks in the mud. Sure, careers have been built on measurement and observation of empty space, but these really involve either measurements of a distinctly indirect nature or converting the empty space into something juicy to observe and measure. Once again, the fact that empty space is unobservable is a double-edged sword.

On one hand, it's highly likely that there would be fewer paradoxes to ponder if empty space were directly observable. On the other hand, our view of the universe would be terrible since pesky, now-unobservable empty space would instead obscure our views in every direction. Of course, on the third hand, the universe would hardly be the place we currently don't know and love since observable empty space would most assuredly guarantee a very different universe than the one we appear to inhabit. In fact, observable empty space would likely mean we wouldn't be around to banter with one another about the nature of the universe.

Chapter 18

The Absurd Unlikelihood of

You Being Here to Read This Book

The truth is that this particular concept has not been lost on scientists and present-day philosophers, some of whom are actually practitioners of hard science. The point is that it would first appear that an incredibly unlikely cosmic coincidence resulted in assembling just the right properties into a single universe to allow life to form and ultimately ask questions about the ultimate purpose of our universe. As some of us are already keenly aware, the answer to the ultimate question of life is "42," as Douglas Adams, now deceased, shared with us in his series begining with *Hitchhiker's Guide to the Galaxy* (New York: Ballantine, 1995). While this may satisfy some of those among us, others, such as myself, are left unsatisfied and must take our quests even further.

Many scientific researchers have developed an acute appreciation for the sheer unlikelihood of the pairing between the physical properties of our universe and the fertile ground it represents with respect to the emergence and sustenance of life, intelligent or otherwise.

Wildly creative theories have been advanced in what appear to me to be fruitless yet deeply earnest attempts to capture the depth of their author's appreciation for this unlikely pairing. I really don't want to denigrate or diminish the attempts at theories to explain these phenomena, because after all their creators crave as much as I do to make some sense of it, as I'm sure many others do as well. This might even be your motivation to read books such as this one. In fact, these individuals each took bold and scary steps to share their thoughts and concepts, many of which may forever remain outside the realm of hard science. Their theories both embody and effectively reveal the deep frustration and sincerity we share in understanding a world that has so carefully guarded its innermost workings and evaded understanding by the very creatures it spawned.

Let's examine some of these theories. As we do, please remember to try to maintain a full appreciation for the motivation behind each theory. It would also probably be a good idea to keep in mind that the difference between matter and empty space is less than one part in over 500 trillion.

Three of these theories seem to stand out a bit more prominently than the others, so let's take a moment to familiarize ourselves with the basic idea or ideas that each embraces. I would certainly encourage the reader to further investigate these and the many other theories not mentioned here, which attempt to explain the incredibly unlikely prospect of a universe that spawns and supports life. Please remember that my attempts at providing succinct descriptions of these theories are unable to capture the fidelity and richness of concepts provided by their creators.

Brandon Carter introduced the "Anthropic Principle" during a talk he gave in 1973 at a symposium honoring the 500th birthday of Copernicus in Krakow. Carter, a theoretical astrophysicist, proposed this term as an alternative to the Copernican Principle, which asserts that humans do not occupy some sort of special, privileged, nepotistic position in the universe.[11]

The basic tenet of the Anthropic Principle is simple even if it first appears to somewhat sidestep the issue. At its basic core, the Anthropic Principle states that the only way we could even be here to discuss any topic whatsoever is because we exist in a universe that is compatible with the emergence of life. Put another way, if the physical properties of the universe were such that they were inhospitable to the emergence of life, we could not be here to hold this dialogue. Therefore, since we are here and having our little tête-à-tête, by consequence, ours happens to be a universe in which life could, and has, emerged even though its likelihood is vanishingly small. Since "vanishingly small" is still infinitely larger than zero, there was some possibility of this particular combination coming into being, and it did, luckily for us. So, here we are.

"Many-worlds" is a concept that emerges from a unique, well-accepted aspect of quantum mechanics, which embraces the objective reality of the universal wave function but denies the reality of wave-function collapse.

Try not to dwell too much on the universally observed perception of the collapse of the indeterminate wave function down to a single observed, determinate state. The logic here gets a bit bumpy, so tighten your seat belt while we temporarily suspend beverage service until things smooth out.

The implication here is that quantum mechanics has decided that all possible alternative observations, beside the one observed, remain real. This, in turn, means that all alternative histories and futures are real, as well as the alternative presents. The consequence of this reasoning, then, is that each of these alternative histories—presents and futures—represents an actual "world" or "universe."

This "many-worlds" concept goes by a variety of names, including the Many Worlds Interpretation (MWI), the relative state formulation, the Everett interpretation, the theory of the universal wave function, the many-universes interpretation, as well as the plain vanilla version—"many worlds."

Many-worlds offers a means by which quantum mechanics can reconcile and perceive apparently non-deterministic events, such as the random decay of a radioactive atom, with its deterministic equations. With the emergence of many-worlds, quantum mechanics no longer views reality as a single unfolding history. Instead, many-worlds views reality as an infinitely branched tree from which every possible quantum outcome occurs.[12]

I think we're ready to resume beverage service now. I know I am.

This theory is also related to the "multiverse" concept. The multiverse, also known by the terms "meta-universe" and "metaverse," is the complete hypothetical set of possible universes (including the historical universe we perceive) that together comprise the totality of reality. This total reality consists of the entirety of space-time and matter-energy (or mass-energy), as well as the physical laws and constants that describe these things.

Perhaps we loosened out seat belts a little prematurely.

The American philosopher William James coined the term "multiverse" back in 1895, apparently to prevent Europe from holding a monopoly on outrageously creative cosmological philosophies.

The multiverse consists of some inconceivable number of universes, each of which is "parallel" to its universe brethren. From here, the number of variations on the multiverse theme just gets insane. Each variation of the multiverse theme proposes some unique set of features for the structure of the multiverse, the nature of each individual universe within it and the relationship between all of the universes within it.

Multiverses are mentioned in cosmology, physics, astronomy, religion, philosophy, transpersonal psychology and, of course, fiction—particularly, science fiction and fantasy. Parallel universes are also often referred to as "alternative universes," "quantum universes," "interpenetrating dimensions," "parallel dimensions," "parallel worlds," "alternative realities," and "alternative timelines," among others.

In case you missed it, hard scientists seriously consider and incorporate the multiverse concept in their non-fiction hard-science fields of cosmology, physics and astronomy, even though science has no capability by which to evaluate the truthfulness or non-truthfulness of the multiverse hypothesis through scientific schemes.[13]

The central theme of these theories is that there is a large number, perhaps even an infinite number, of parallel universes that coexist simultaneously and undetectably. Each of these universes varies in some sort of way. Some of the differences among them are large and some small, but in each case some fundamental physical property, or set of physical properties, take on values that are different and unique for each universe. In other words, the set of values for each of the fundamental constants in each universe is unique with respect to the set of values for each of the fundamental constants possessed by each of the other universes.

Among those many universes are a large number of universes that are unfavorable to the emergence of life, and a very, very small number of universes in which the conditions are favorable to life. In such a scenario, the probability of any particular universe harboring life remains vanishingly small, but that small probability is multiplied by a vast number of universes. The end result is that, as unlikely as the right conditions to spawn life might be, a small number of universes with all the right stuff can, and does, form. Ultimately, we can hold this dialogue only because we had the good fortune to be in the right universe, or one of a very few favorable to life.

There's another variant of the many worlds theory that bears some resemblance. They vary in that the universes form in series. First, one universe forms in its own personal Big Bang and finally undergoes its own personal Big Crunch, resulting in this universe's demise and the starting point for the Big Bang of the next generation's universe. So, over time—a very, very, very long amount of time—a very large number of universes ultimately form, one after the next.

Furthermore, with each collapse and rebirth, the subsequent universe takes on properties subtly different from those of its predecessors, as the energy of each universe diminishes incrementally with each passing cycle. As the cycle of universe formation and collapse unfolds, the sum total of incrementally different universes grows and grows over time, and a very small percentage of the formed universes have precisely the right values for life. In this scenario, we have the good fortune to be in one of the small number of universes that emerge over time that are favorable to life, and we can therefore engage in our dialogue.

The last of these theories that I've chosen to include here is also among the newest. The Theory of Biocentrism was proposed in 2009 by Dr. Robert Lanza. "Robert Lanza is an American Doctor of Medicine, scientist, Chief Scientific Officer of Advanced Cell Technology, and Adjunct Professor at Wake Forest University School of Medicine."[14]

Biocentrism is unique among the theories I've included here, because it removes the element of probability inherent in each of the other theories and replaces it with an intimate relationship between the universe and life that is strongly symbiotic in nature. The essence of this theory is that the pairing of favorable conditions and the presence of life is not a coincidence whatsoever. The two elements—those being a universe possessing the right fundamental physical properties to support life and life itself—are intricately intertwined and inseparable. I hope that I'm not putting words into Dr. Lanza's mouth, but it might not be inaccurate to describe biocentrism as supporting the notion that life itself might even play some role in ensuring that the universe possesses the right attributes to support its very existence. Thus, the combination of a hospitable universe and the emergence of life being highly unlikely is itself an illusion and the right conditions are not left to chance at all.

With *Biocentrism* I felt I finally understood the significance of the assertion that space and time are properties shared between the observer and the universe. The concepts presented by Dr. Lanza had me thinking in ways I hadn't previously considered. They provided me with a new framework of thought by which to re-interpret so many scientific experiments and paradoxes.

I decided to let new thoughts and perceptions flow and see just where it would take me.

I think the entire scientific community is pretty much in agreement regarding the outrageous unlikelihood of the universe possessing precisely the right physical properties that would result in precisely the right conditions for life to emerge.

I certainly thank my lucky stars that what unfolded in our universe led to my existence. Now, I don't know about you, the reader, but I never hold the winning ticket for any sort of door prize or lottery. (No, I don't buy lottery tickets! What would be the sake?) I'm convinced that if there were only two tickets for any door prizes at events I've attended and I was holding one of them and I went to a hundred events, I'd still never be holding the winning ticket!

This got me thinking....

Chapter 19

I'm Not Lucky

I'm not at all a superstitious person, and I believe in probability, not in luck. Having said that, I feel that probabilities don't have a tendency to work in my favor. (Remember, I'm not talking about luck; I'm only relating objective personal observations to you.) The probability of a universe possessing all the right characteristics for the emergence of life *and* my being present in this universe seems to me to be an inconceivably unlikely coincidence. As many of you might already believe, there are no coincidences.

So, the prospect proposed by Dr. Lanza that life emerged neither from some cosmic coincidence nor from an inconceivable number of failed universes, either in series or in parallel, became more and more appealing to me.

Of course, it's possible that none of the theories mentioned above have any validity to them whatsoever and the theory that accurately explains the hospitable nature of our universe and the consequential emergence of life may still be waiting for someone to think of and propose, but at any moment in time we can only select from the options that are available to us. Yet the concept of countless dead parallel and serial universes seemed to convey to me interpretations that were the very opposite of those resulting from consideration of Lanza's concept. The very prospect that I'm here to hold this discussion because of a coincidence is a concept that I am loathe to consider. And the argument offered by the Anthropic Principle, at least to me, seems pointless and dead-ended.

In many ways the Anthropic Principle bears some resemblance to the famous statement made by Descartes: "I think, therefore, I am." In a sense the Anthropic Principle seems like a restatement of this concept, something along the lines of

"I am here; therefore, it is necessary that I am." But I also feel that the Anthropic Principle can readily be distinguished from Descartes' statement by turning it upside down, such as in "There's no way I could be holding this discussion if I didn't exist." This is an obvious statement of truth at the core of which is very little meaning. If I didn't exist, then the only statement I could ever utter would be " " and this is equally meaningless.

Furthermore, the entire concept of one or more Big Bangs representing the start of our universe, or anyone else's, and the moment at which time began its flow seems awfully misplaced as well. In my opinion, more scientific paradoxes arise from the concept of the Big Bang than from any other single scientific theory, which indicates to me some sort of misunderstanding. The Big Bang may very well be the "mother of all misunderstandings." But that's a discussion left to a later chapter.

Let me remind you of the other statement I could make that also begins with "of course" and that is this: "Of course, I could also be entirely wrong." So, you'll have to make your own judgments about the path I've chosen to share with you. Hopefully I'm completely correct in every assertion and hypothesis I propose, but that could be viewed as a coincidence, since much of what I propose doesn't have a complete and thorough scientific basis. Yet, even if I'm wrong about everything I've chosen to include here, I remain hopeful that I've succeeded in introducing you to new ways of thinking and new templates by which to interpret the universe around us. In this case, the expanded thinking process that might result from reading this book might provide the seed for one of its readers to develop the concepts that provide the remaining missing puzzle pieces to humanity. (Either way, I win, so there you have it.)

Still, the very concept of a universe characterized by symbiotic relationships with its observer inhabitants is very much along the lines of what we've already been discussing throughout this book. It is becoming a stronger and stronger prospect, in my mind, that to attempt to refute this symbiotic aspect of our universe is nothing more than the ultimate exercise in futility. I hope to convince you by the end of this book that resistance is useless. (What? You didn't really think I was going to resort to an overused cliché from popular science fiction, did you?)

So with all of these cumbersome arguments in mind, let's see if we can make heads or tails out of the biggest mystery that humanity has yet encountered. No, I'm not referring to Paris Hilton; I'm talking about the void. We need to crack the nut presented by the concept of nothingness before we can solve the rest of the universe's mysteries.

Chapter 20

Understanding the Poor, Neglected Void

As I've already mentioned, void is the single biggest component of our universe. The common perspective in science is more along the lines of "the void is nothing" so there's not much to discuss. From my perspective, however, the void is not only the biggest component of our universe, it's the most important.

As usual, let's try to summarize where we are before we proceed.

Beginning about 2,500 years ago, philosophers accurately theorized that matter was comprised of tiny fragments, which they called atoms. The "atoms" of these pre-Socratic philosophers, represented the most fundamental building blocks from which the matter they observed could be constructed. This school of philosophy became aptly known as Atomism. It would not be until the twentieth century that the scientific method resulting in large part from the philosophy of Atomism could itself be used to confirm their theories. The atoms of the Atomists describe the most fundamental particles from which matter could be constructed—unlike the modern use of the word atom, which describes the basic chemical unit of an element, comprised of particles we call electrons, protons, and neutrons. To the best of scientific understanding today, the electrons are fundamental particles that cannot be broken down any further, while protons and neutrons are themselves comprised of more fundamental particles called quarks held together by quanta of energy, appropriately called gluons.

Science focused largely on matter, which is only one of the two fundamental parts of the universe that Atomists like Parmenides proposed. The other fundamental part of the universe is void, the concept of which is difficult to embrace in English. Parmenides summed up his thoughts about the void in a very succinct statement, "Being is, while nothing is not," which once again loses its effectiveness since he said it in Greek and we're reading it in English.[15] By way of summary, Parmenides felt that the void is the ultimate non-existence and that it's inappropriate to describe it using any verb which in and of itself conveys existence, like through the unavoidable use of conjugations of the infinitive verbs "to be" or "to exist." But the really interesting part of Parmenides' propositions, in my opinion, is his point regarding ultimately fundamental "atoms" of matter, which could not be broken down any further. He used a very illuminating, insightful description. He described an "atom" as the basic component of matter which could not be split any further by the *void*. Herein lies a really juicy description of the void into which we can sink our teeth.

Parmenides' concept becomes particularly interesting when we recall that a carbon atom consists of almost nothing more than empty space—less than one part matter by volume out of more than 500 trillion is comprised of what we consider today to be fundamental particles, those being electrons and quarks.

When we think of larger objects, which contain open spaces within their designs, we really don't spend much time thinking about the nature of the open space (like a balloon or a beach ball), because we understand that if the balloon or beach ball is firm and bouncy, and not flat, that the empty spaces are filled with air—a mixture of gases composed primarily of oxygen and nitrogen molecules. So, we go about our lives in our daily macroscopic existence pretty much apathetic about volumes of space that appear totally empty, since we are unable to "see" any filler but which we know *intellectually* to be chock full of gas molecules rushing this way and that. We get pretty used to the concept and really don't think much about it.

But let's delve back into the structure of the atom. We have a nucleus comprised of fundamental particles called quarks and quanta of energy called gluons. We have a shell of electrons that flit throughout some volume of space permitted them by the probability wave equations that quantum mechanics calculates for them. And throughout the comparatively cavernous volume of the atom defined by the vague outer edges of its electron shells is apparently empty space in between each of the fundamental particles. And lots of it, as we've discussed. The "ball" of the atom is not inflated with gas molecules. There isn't anything that we consider "physical" that fills the cavernous open spaces. There's nothing smaller than the fundamental particles that could be filling the space. There simply isn't anything macroscopic with which to make a meaningful analogy. Of what could the empty space possibly consist?

Well, perhaps we should begin with what the empty space itself does not contain. If we divide the world in two, along the lines of what Parmenides suggested, there is only matter and void. We know that we have excluded the volume of matter from our calculations of the empty space within the carbon atom, so let's just zero out the vanishingly small portion of matter for the moment.

Einstein's relativity showed us that matter and energy are equivalent and can be converted from one to the other according to the famous equation, $E = mc^2$. So, if we remove just the matter, we're really also removing some amount of energy. In other words, if we eliminate the matter comprising each electron, then we're also removing the electromagnetic force that each electron brought with it to the party.

As we also discussed, electromagnetic force is carried by photons, quanta of energy, which travel through empty space as indeterminate waves until our observations cause them to collapse to determinate states that manifest themselves as the "particle" faces put forward by photons. The positively charged nucleus attracts the oppositely charged electrons, while the electrons repel each other, since they possess the same negative charge as one another. The gluons, too, are quanta of energy that keep the different flavors of quarks organized into protons and neutrons. The strong and weak nuclear forces are also comprised of energy quanta that keep the cluster of neutrons and protons together in the nucleus and overcome the electromagnetic quanta from the positively charged protons and keep the apathetically uncharged neutrons from leaving the party.

All of this should provide the impression that the empty space is not so empty. It may seem physically empty, but there are quanta of energy moving about the insides of the atom in a flurry of activity to preserve the structure of the atom. So, the empty space is not completely empty, after all. This is the conclusion of researchers who study the physics of "nothingness" in an attempt to understand the nature of what science refers to as "free space."

There are a couple of reasons why I find this answer flawed and incomplete. First, if you sum up all of the quanta and derive an equivalent amount of matter, and then determine the volume of that matter, you still end up with way too much empty space. But, more importantly, upon what exactly are the indeterminate probability waves of the quanta traveling? After all, from our experience in the macroscopic world, we expect waves to travel through a medium. Acoustic waves travel through the air, alternately compressing and expanding the density of air molecules along their path. Drop a stone in a pond and watch the mesmerizing waves forming and spreading along the surface of the water. Is there some meaningful analogy for the way that waves of strong, weak, or electromagnetic forces propagate?

Surely Einstein or others got to the bottom of this and a definitive resolution exists, right? Right?

Well, there's some debate about this, but I contend that the answer is no, not really. Einstein's relativity is quite specific in describing the way in which electromagnetic force travels through free space, but his theory doesn't define the internal nature of free space. Along similar lines, the Vikings knew how to design and build fantastic sailing ships without ever knowing the fundamental composition of the water upon which their ships travelled.

Yet, relativity effectively ended a divisive dispute within the scientific community that lasted for decades preceding its emergence. The dispute involved precisely the topic of how light travels through space. Some scientists supported the concept of the Luminiferous Aether, a mythical medium upon which light could travel. Hence the "Luminiferous" Aether carried "light" waves and waves of the entire electromagnetic spectrum.

Let's review some of the debate.

By 1864 Maxwell had completed the development of the final form of his equations, which describe the relationships between electric fields and the intertwined magnetic fields they generate, and vice versa. These equations definitively unified electric fields and magnetic fields once and for all into a combined electromagnetic field. These equations also definitively established the relationship between electromagnetic waves and the phenomena known as visible light, and created the basis for our understanding of electromagnetism and the electromagnetic spectrum in place today.

Einstein's thought experiments permitted him to combine Maxwell's equations with those of Lorentz into a single context he called relativity. Relativity describes the behavior of light when viewed from differing perspectives. Relativity also describes the ways in which matter's mass curves empty space to create gravity. In fact, relativity theory covers quite a lot of stuff. Relativity shows us the equivalence of mass and energy and it shows us the equivalence of gravity and acceleration.

The debate surrounding the Luminiferous Aether became most heated in the time between the publication of Maxwell's equations in 1864 and Einstein's Special Theory of Relativity in 1905.

Perhaps the tension in the scientific community climaxed with the Michelson-Morley experiment, the results of which were published in 1887. Several related experiments were performed by other researchers prior to and subsequent to the Michelson-Morley experiment, but the Michelson-Morley experiment set the standard to which these other experiments were compared because of the precision with which the experiment was performed.

Essentially, the concept behind the Luminiferous Aether was simple. Everything from the perspective of our daily experiences just screams at us that waves have to travel on or through something, yet the void of free space represented the absence of something, or as Parmenides reminds us, "nothing is not."

Numerous scientists proposed hypotheses for a medium called the

Luminiferous Aether upon and through which electromagnetic waves travel, including Maxwell and Lorentz. The hypotheses described in detail the properties that this Aether should possess; the Aether was hypothesized to pervade our universe from one end to the other. With respect to an observer on our Earth, which is simultaneously rotating on its axis, revolving around our sun, and moving through our region of the galaxy that in turn is moving through the space of our universe, it was further hypothesized that the Aether could either be stationary or it could be moving. The hypotheses went on to propose that an observer would measure different values for the speed of light depending upon which way the Aether was "blowing."

Let's try to imagine a situation based on everyday experience that might provide a useful analogy. Imagine throwing a baseball while standing still and a radar gun placed nearby determines that you threw the baseball at 100 miles per hour. Then imagine standing up out of the sunroof of a car travelling at 100 miles per hour and then throwing the baseball in an identical fashion to the way you threw it while standing on the ground. Discounting any change in wind resistance between the two circumstances, the baseball travels away from you at the same speed of 100 miles per hour. It doesn't matter to you whether your feet are planted on the ground or on the floor of the moving car, the ball still travelled away from you at the rate of 100 mph. But, to an observer equipped with a measurement apparatus who is standing stationary on the ground, he would measure very different values for the two situations. In the first case, the observer measures the baseball travelling at a speed of 100 mph. In the latter situation, however, he measures the baseball moving at a speed of 200 mph, which is the algebraic sum of the 100 mph speed at which you threw the baseball and the 100 mph speed at which the car was travelling when you threw the ball.

It was reasoned, by analogy, that a similar thing would be observed regarding light and the motion of the Earth through the Aether, based upon every experience that humanity had accumulated up to that time. In other words, the speed of light would be observed to be different in the direction of the Earth's motion through the Aether than it would be in the direction orthogonal to Earth's motion.

Albert Michelson and Edward Morley designed and built a sophisticated apparatus that would permit them to measure and compare the speed of light observed simultaneously along two orthogonal paths. The experiment was performed and measurements were taken. The subsequent analysis showed that the speed of light was identical in each direction.

Since this violated common sense based on experience up to that point in time, researchers had a difficult time believing the results, and variants of the experiment were performed over and over, producing the same negative results.

In 1905 Einstein published his Special Theory of Relativity, which was heavily based on prior work by Lorentz, Maxwell, and Henri Poincaré. Einstein's Special Theory of Relativity, and the calculations he performed, showed that our prior everyday perceptions of the universe were deceptive and that the speed of light was the same for each and every observer regardless of their particular perspective or circumstances. Following the emergence of Einstein's theory and arguments, the concept of the Aether was largely abandoned. Ten years later, in 1915, Einstein published his General Theory of Relativity, which concerned the equivalence of gravity and acceleration.

An interesting item for us to note here is that although Einstein's Special Theory of Relativity spoke quite a bit about free space and accurately described the rules with which electromagnetic waves propagate through free space, Einstein didn't attempt to delve into and describe the fundamental nature or structure of free space. The observation of light's behavior through space provided a great deal of insight about our universe, at least from the macroscopic point of view, but only provided indirect evidence about the most fundamental aspects of the void. As we've learned, indirect observation of the void may be as good as it gets, after all.

In the big picture, though, indirect evidence of unobservable objects and phenomena is very different from having no information at all. In fact, it's much better than nothing. If we can assemble and assimilate the right pieces of perception, interpretation, and understanding, it might be possible to construct a vividly detailed portrait of any object or phenomenon.

In the macroscopic world of our daily lives, we can find ample examples of analogous situations. Forensic portrait artists assemble fragmented bits of information from witnesses and victims of crime to create a composite image used to capture alleged criminals without ever having observed the perpetrators themselves. People who like to assemble jigsaw pieces also deal with analogous situations as a matter of course. These very patient and methodical people seek one particular piece at a time often knowing nothing more than the shape of its absence. In effect, these puzzle people are able to construct a detailed image in their minds of the piece they seek based only on the negative space defined by its absence. The missing piece appears as nothing more than a hole in their puzzle, and armed with the image it creates in their minds, they set out to find it. They know precisely when they've found it because it completes an image which, as we've learned, exists nowhere else in the universe than in their minds.

In fact, the more you think about it, the more familiar this sort of situation is in our daily lives. More often than not, we have an incomplete set of information and somebody is expecting us to make some sort of decision based on the limited information we actually possess. But, oftentimes, we actually have two types of information, even though we might not always notice it in our frustration. As we just mentioned, we have some information of the type that is known and available, and we also have the sort of information that hasn't been provided to us. Even though we don't have all of the information we desire, we have more information than we consciously acknowledge, since in the vast majority of circumstances we know *which* information is missing. That, in and of itself, is information. It may not be in the form we prefer, but it's information all the same. This type of information is analogous to the limited, negative-space information that puzzlers often use as their primary tool to solve a puzzle, no matter how complex it is, like perhaps a puzzle involving an Escher drawing.

When push comes to shove, the vast majority of situations in our lives are like this, and we still make our decisions and plow through problems and challenges with astonishing success for which we probably take little notice or credit.

I propose extending this analogy to the subject of the void, and then we'll deal with the consequences.

So what do we know or think we know? We know that electromagnetic waves travel through space with apparent ease, although something about free space limits light's observed speed to roughly 300 million meters per second. We know that we can look right through empty space to observe objects and phenomena near and distant. We know that virtually all of the volume of an atom is comprised of empty space, and by extension, that all matter consists of mostly empty space. Put another way, we know that empty space is what separates each fundamental particle from another, a concept first introduced by Parmenides and the ancient Atomists. We know that free space is not directly observable but that anything we do observe comes to us only after quanta of energy emerge from the void and kindly collapse from an indeterminate quantum probability wave down to a phenomenon we can then observe and measure. We know that the results of the single-photon, double-slit experiments indicate that the presence and interaction among these indeterminate probability waves creates the perception of an interference pattern that we observe. We know that mathematical physicists have described the appearance of the resulting interference pattern by a mathematical analogy embodied in Feynman's Sum-Over-Paths Model.

It stands to reason, therefore, at least in my mind with the interpretations I've embraced along my way, that the void is comprised of the total sum of individual indeterminate probability waves, both the small number that result in observations of objects and phenomena and the much larger number of indeterminate probability waves that don't. The sum of all the individual indeterminate probability waves is represented by a composite indeterminate field that represents the large but finite potential energy of the universe from which all observable objects and phenomena emerge.

Ultimately, the full range of possible observations that can be perceived by an observer occur as a consequence of the interaction between the observer and ɘr internal or external, or both. I'm proposing that the very act ⁣uses some small portion of the universe's total potential energy, ⁣ɪe quanta that we've come to know and love, to collapse to an observable state we deem as matter (object) or energy (phenomenon). That's right. I'm saying that matter and energy are ultimately different manifestations of the potential energy of the universe and are, therefore, equivalent to one another in a way that constructs a self-consistent bridge between relativity and quantum mechanics.

Here's my hypothesis, then, regarding the void. We will refine and add to this hypothesis by expanding the concept and definition of the void step by step. As we'll also see, numerous other scientific hypotheses and mathematical constructs will be folded into our definition as we expose these concepts to be nothing more than synonyms for the concept of the void that will soon be included in our growing glossary.

I'll go one step further. Whereas we perceive matter as being organized from individual, discrete, indivisible, fundamental particles and energy manifesting itself in the form of four different forces—which project their forces through their corresponding bosonic carriers in the form of quanta—the void is topologically one contiguous entity acting as the underlying substrate or connective tissue upon which mass and energy perform their dances to our observation, entertainment, and befuddlement.

We're all fond of finding ways to divide the world into two segments as we've discussed. Examples include "us" versus "them," "mind" versus "body," "hard" science versus "soft" science, "science" versus "religion," "science" versus "spirituality," "internal" universe versus "external" universe, "mass" versus "matter," and "matter" versus "void." In each case that we've examined thus far, we discovered as we bore down from our "perceptions" to a better understanding of the situation through interpretation and assimilation of concepts into a bigger picture through abstract reasoning that the very notion of a boundary was artificial, synthetic, or unreal. Although it might come off as a bit spiritual, the interpretations we've been constructing led to a couple of inevitable conclusions.

The first irrefutable conclusion is that it is the nature of humankind to find ways to simplify and compartmentalize our big, complex world into smaller, bite-sized pieces with which we can better deal. The simplest and easiest way to do this is to iteratively divide the world into two pieces: the "this" part and the "that" part. This human tendency seems universal, a sort of innate behavior inherent in humans and the human condition. This element of our identity is reflected throughout the history and geography of human culture. An almost inconceivable number of people have perished as result of the simple assertion that the world can be divided into two pieces labeled simply "mine" and "not mine." You likely could provide ample examples of this human attribute from your own experiences and circumstances. It would even seem to me that this particular simple human trait throws into question whether humanity will succeed in overcoming it and surviving, or continuing to live with the vile consequences of violent behaviors and perish.

Nonetheless, this all-too-human characteristic insinuates itself into every aspect of our lives, culture, and society. It seems to be everywhere I look for it. And among the myriad places where this phenomenon can be found are within the tools and concepts emerging from science. Its presence in science and the mathematical constructs it has developed and continues to use to frame and answer questions isn't anything special, as I've noted, since the tendency is universal in its extent throughout human culture. The only thing that makes it distinctive with regard to science is that our expectation for science is to illuminate the real, innermost workings of our universe and to share its finding in a positive, constructive way. In so doing, we serve and enlighten humanity so it can hopefully realize a destiny ahead that improves upon the past struggles characterized by viciousness and in-fighting. If people can come to truly understand that boundaries don't really exist in the fabric of our cosmos, perhaps we can finally stop trying to build them.

Science's mathematical tools and the images we produce through our increasing technological sophistication reduce the reality of unbiased, methodical measurement to a false-colored, but enormously simplified, interpretation of our world. The competition among scientists to produce pretty pictures to support their funding requests and the stature of their organizations may very well distract us from the ensuing stagnancy in better understanding our universe even as these stunning pictures positively dazzle and perplex the public, in the bid for scientists to secure funds from a static or shrinking pool while the costs of scientific experimentation soar dramatically upward.

I really don't mean for this to come across as "evil" as my description might have sounded. It seems to me to be the nature of humanity and human culture to go through these fits and spurts. Science enjoyed outrageous, electrifying results during the twentieth century. In all honesty, it's a tough act to follow. Human history is fraught with examples that are analogous to those of twentieth-century science, as innovation and human energy build upon one another in the frantic and fruitless attempt to grow and expand at a rate that is simply not sustainable. Sure, external factors like exponential cost growth and resource depletion can profoundly limit growth. Yet a more significant factor in my mind is that successful cultures ultimately first consolidate their gains and enjoy the fruits of their labors and then they become complacent about continuing innovation, while displaying hostility toward those willing to expose the lamentable wave of decline that is beginning to well up.

In essence, the tools, models, and limited understanding science has achieved all have the potential to become a sort of prison that limits its very own advancement.

Anyway, the second irrefutable conclusion is that there are no such things as boundaries as we delve most deeply into the finest granularity of our universe's fabric. The boundaries we perceive in our universe are illusions arising from the combination of the nature of our perceptions and our propensity to divide the world into convenient pieces.

Here's our next glossary term:

◊ **Void** – a single phenomenon comprised of the collective, composite, indeterminate probability field representing the full potential energy of our universe, for which there is no direct observational relationship to a particular observer or set of observers, and which is therefore undetectable and unobservable.

Consider this the hypothesis of the void with which we are now working. We'll refine this definition later.

Let's begin to break down this concept of the void. First, despite my diatribe, I am human and I can't be anything other than human, so my hypothesis breaks the universe down into two parts that can be described by the terms "potential energy" and "not potential energy." The former term represents the void, while the latter term refers to all of the matter and energy that we can potentially observe through direct means and with which we are at least somewhat familiar. This latter group includes any manifestation of the electromagnetic force and its bosonic carrier, the photon. I'll refer to this latter group as "condensates" resulting from the collapse of probability waves from the former group into potentially observable, directly detectable objects and phenomenon. The collapse of a quantum from the composite indeterminate probability field, or "void," results in the coalescence of a manifestation that we perceive as an object or phenomena within the mass-energy domain to which we are familiar and which is represented by both Newton's laws of motion and Einstein's relativity equations. The coalescence of the void into one of these manifestations reduces the local potential of the void, and provides observers with measureable phenomena. I contend that the measureable phenomena can be divided into direct observation of luminous matter via electromagnetism and indirect observations of the void, which again is the singular, composite potential energy of the universe.

Let me try to explain this in terms designed to reduce the complexity and mystery down to more graspable language. Everything we see and otherwise observe is what we perceive to be actual and real. These objects and phenomena would appear to be "created" out of the thinnest air possible, the "void," through an intimate interaction between observer and the composite potential energy of the universe, also said "void." The void, unlike our perceptions of photons and particles of matter that appear to us as discrete, separate things, is a single, topologically contiguous phenomenon that is directly undetectable and unobservable to us. The act of observation produces the familiar objects (particles), which we can then observe and measure to our heart's delight.

This definition allows us to view many other related scientific concepts and models as synonyms of "void." These include familiar terms like vacuum, space, empty space, and free space, and terms that may not yet be familiar like vacuum fluctuation energy, quantum fluctuation energy, chaos, and the Dirac Sea. It may also be appropriate for me to include the term dark energy here, as well, as I will explain later.

Chapter 21

Assembling the Pieces of the Puzzle

Let's keep the universe simple and boil my hypothesis down to the simplest, most succinct statement and keep the most important elements intact.

The universe consists of energy in two forms: potential energy and "kinetic" mass-energy, for lack of a better term. Perhaps I will create a new term to better capture the concept of "observable" mass-energy. The potential energy of the universe is not in a form that we can directly observe. The potential energy seems like a somewhat nebulous concept to us, since it remains forever elusive and directly unobservable, and unfortunately there's nothing science can do to change that. "Observable" mass-energy takes the form of photons or objects that we perceive as matter comprised of vanishingly small fundamental particles.

As a further clarification, let's take a peek back at the representation of the Standard Model. The carriers of force are the bosons, namely the photon, gluon, Z-boson, and the W-boson. The particles of matter fall into two subgroups. The leptons with electrical charge states of -1, 0, and +1 include the very familiar electron, and the likely less familiar muon and tau particles along with their corresponding neutrinos. The other group consists of the quarks with their interesting and colorful names. Together the leptons and the quarks comprise the larger classification of fermions, which are organized into three generations of matter represented by their respective "charge -1" particle, the electron, the muon, or the tau particle.

The structure I propose, namely "potential" energy versus "kinetic" mass-energy divides the world into two segments in a way with which both science and we are already familiar.

In our everyday lives, we're familiar with the notion that an object in Earth's gravitational field possesses varying amounts of potential and kinetic energy. For instance, if a Hummel figurine is sitting on a shelf above the floor, then the figurine possesses a certain potential energy value that can be calculated using the equation, $PE = mgh$. If we drop the Hummel figurine to the hard, unforgiving floor, the figurine's potential energy converts to kinetic energy as it falls and the kinetic energy it possesses as it shatters into pieces is calculated by using the equation, $KE = \frac{1}{2}mv^2$.

The potential energy possessed by the figurine sitting on the shelf is precisely the same value as the kinetic energy it possessed at the bottom of its fall. When the figurine is sitting on its shelf, it possessed no obvious potential or kinetic energy, since it wasn't moving, yet its energy was stored as unobservable potential energy. Conversely, as it fell, the figurine's potential energy was converted increasingly into kinetic energy until the point when it no longer possessed any potential energy and its energy was converted into kinetic energy, which happened immediately before the sound of shattered ceramic.

By the way, the simple description I provided a few paragraphs above of the universe consisting of potential energy and "kinetic" energy represents an example of a concept that has come to be known as Ockham's (or Occam's) Razor.

William of Ockham (c. 1288–1348) was an English Franciscan friar and scholastic philosopher presumably born in a small village in Surrey. Many believe that Occam, regarded as quite the controversial figure in his time, both politically and intellectually, greatly influenced medieval thought. He made major contributions to the fields of logic, physics and theology, yet he may be best known for the principle that bears him name: Ockham's Razor.[16]

William of Ockham believed in getting to the root or essence of a concept or issue by reducing it down to its shortest possible description—a description that still faithfully preserves the essential elements and accurately preserves the essence of the idea. He also stressed the prospect that you could be overzealous in the use of your razor. In this case, you might cut too much material out, which would result in missing key elements that then need to be added back in. The basic idea behind Ockham's Razor came to mean something more along the lines of "the simplest explanation is usually the correct one."

Either way, Ockham's Razor is a term that's used quite a bit in the science community as the plethora of facts and collected data need to be pared down to prevent a simple explanation from being obscured.

Einstein certainly captured the essence of Ockham's Razor when he said, "Everything should be made as simple as possible, but not simpler."

So let's examine the Ockham's Razor version of my hypothesis, namely that

II: Assembling the Pieces of the Puzzle ❈ 165

the universe is comprised of potential energy and "kinetic" energy in the form of mass-energy particles.

By this, I mean that the sum total of energy in our universe is constant. Sure, the energy distribution can be shifted between kinetic and potential but, as a closed system, energy can neither be lost from the universe nor can the universe magically acquire new energy. In essence, the energy of our universe represents a "zero-sum game," in which the sum remains constant under any and all conditions. And, by my use of the term "constant," I specifically mean a huge value but a finite value, nonetheless.

Remember as I discussed earlier, I don't observe any infinite lines or planes in the real world, with the exception of their portrayal in the field of geometry. So, I refute the very concept of infinity when applied to our universe. Once again, the concept of infinity exists in only one place in the universe and that's in our minds. The sum total amount of energy might be positively ginormous but it's not infinite. It fact, the association between infinity and the other explanations attempting to explain our presence in a habitable universe (other than biocentrism) that we discussed earlier left me with rather bad tastes in my mouth. I feel that in many ways, the concept of infinity belongs more in the realm of religion than the realm of science, but these two realms often seem to have a lot in common anyway. As I will share with you, cosmologists often introduce the concept of infinity when they examine the void. Like other attempts to apply the concept of infinity to the real world, I think this one is also misplaced.

Let's go through another exercise. For this one, let's create in our minds an image of some small volume of space that we're going to study in the lab. Imagine we've placed this small volume of space in our measurement apparatus and we're now going to attempt measurements on it. Well, if we don't want any matter that's present in our sample to get in the way of our observations of the empty space in which these particles are immersed, then our first order of business is to eliminate all of the matter in our precious sample of space, so we attach pumps to our measurement apparatus to suck the molecules of air out of our sample. In reality, our pump technology is really quite effective at sucking molecules out of the space, but there are always some uncooperative molecules of air that remain. These molecules take refuge on the inside surfaces of our vacuum chamber and are very difficult to eliminate completely. So, in reality, our technology allows us to draw a darn good vacuum, but not a perfect vacuum. Since this experiment is being performed largely in our minds, let's just go for it and say we've sucked every last morsel of gas out of our apparatus that was present at the start of the experiment.

Okay, we've eliminated the matter. What's left? We've got the four fundamental forces and their carrier bosons. That won't stand! We'll need to get rid of these intruders before we proceed, won't we?

Let's take the process of eliminating forces and bosons from our apparatus one step at a time. Let's begin with the strong force present in the nuclei along the surface of our laboratory-grade, shiny, steel, vacuum chamber. It's true that the precise position of the vessel wall is difficult to pin down at the quantum scale, but our piece of space is wide enough in each direction that this force doesn't extend into the volume in any meaningful way. Remember that if the space we're studying is larger than anything as tiny as the diameter of an atomic nucleus, the strong force can't effectively exert its influence in our space. A similar situation holds for the weak force, so we can justifiably neglect the effects from these intruders.

Now we come to the electromagnetic force. Aside from the electron shells belonging to the atoms lining the surface of our chamber, we know how to shield any stray electromagnetic fields that might affect our measurements. We've taken the necessary precautions to shield our apparatus from photons straying into our experiment from outside the chamber. In a conductive metal, like steel, the outermost electrons are actually shared among the atoms in the material and become the community property of the atoms in the metal. We can't do much about the presence of these electrons and the fields they produce, but the effect is negligible if our space is sized appropriately.

In essence, we use a little bit of the space in our experiment to provide effective shielding against the three big, strong forces.

Only gravity remains. We know of no means today by which we can shield our experiment from the Einsteinian curvature of space that results from the mass of surrounding matter. The force may be a weakling, but there's quite a bit of mass both nearby (our planet) and throughout our universe. We'll just have to learn to live with it.

Let's review what we've done so far. We've eliminated the matter we can and decided to zero out any matter that remained in the space we're about to observe. We've eliminated any concerns about the strong and weak forces. We've eliminated as much of the electromagnetic force as we can, but we can't entirely eliminate the electromagnetism within our space so the value of the electromagnetic field is very small but finite at any point. The field value is just a tad over zero near the chamber walls. Then, there's gravity, about which we can't do anything, but which fortunately is small in value. If it were possible to place a small gravitational measurement device inside the small space of our vacuum chamber, we would measure the same value for the universal gravitational constant there as we would measure anywhere else.

After all of our efforts, we realize that we can't create a perfect vacuum. We can get really, really close but we can't thoroughly and completely create a perfect vacuum, not even in our mind experiment. And, if we can't even do that in our minds, then it's surely not possible in the real world. Anyway, let's go take some measurements and see whether we can detect anything interesting.

If you're paying attention, this suggestion might perhaps seem like a trick suggestion. After all, hadn't I included in my hypothesis that the void itself is directly unobservable, not measurable, and undetectable? Hopefully, you're awake and alert and following right along with me on this.

We turn on our fermion and boson detectors and let them run for a while. We analyze our data and to our surprise (hopefully mock surprise by now) we discover that we have perceived the perception of mass-energy in the form of fermions and bosons, except they look a bit pale for the most part. If we're patient, these ghostly particles, referred to by scientists as *virtual particles*, come and go. They can also transcend into the real things, like photons or quarks. That's right: real photons and quarks just dropping out of the void!

Your first response has got to be, "No way! Are you making this stuff up?" I can assure you that I'm not making it up; this phenomenon is both real and explainable, even if your science instructors had never bothered to tell you about it. This is the point at which I'll now introduce the scientific concept of the vacuum fluctuation energy, which is also known in science as quantum fluctuation energy.

But, before I do that, I'm going to make a short departure before we go any further and talk about the process of evaporation that we encounter in our everyday lives. I'm a coffee drinker, but the story works just fine if you're a tea drinker. I've entered my middle age and one of the most vivid reminders of achieving that status in my case involves evaporation.

I use a tea kettle to boil water and my wife removed its whistle some time ago and discarded it because she found its sound too unpleasant and shrill. I fill the kettle with water, place it on the hot stove, and walk away. So, when my middle-aged brain finally remembers that I was heating up water for my coffee, I run back to the kitchen to discover that there's no longer enough hot water in the kettle to fill my 18-ounce Bodum coffee press. Then I pour more water into the kettle and repeat. Eventually, I end up enjoying my cup of coffee because I'm persistent, but not before causing my utility bill to needlessly bloat. Recently, I installed a hot-and-cold beverage center at my home because I just couldn't stomach the impact my coffee drinking was having on the environment.

In any event, this wasn't supposed to be a story about middle age and its resulting impact on my carbon footprint; it was supposed to be a story about evaporation. Evaporation is due to a quantum effect whose impact is readily observable at the macroscopic scale.

Here's how it works. You place water at room temperature, or so, into your kettle. By the term "room temperature," I mean that the water molecules have an average temperature corresponding to what we refer to as room temperature, about 68°F. By an average temperature close to room temperature I mean that the water molecules have an average kinetic energy corresponding to room temperature. You see, there are an awful lot of water molecules in the kettle when I place it on the stove and like other objects in nature that we've discussed, these molecules possess a broad spectrum of kinetic energy values.

As they bounce and jostle each other like commuters on a rush-hour train, they constantly transfer energy from their motion to one another in a random fashion. The temperature of an object is related to its average kinetic energy. So, the jostling results in some molecules possessing "temperatures" a bit below the average temperature and some molecules possessing temperatures far above the average temperature. These few but highly energetic molecules may possess a "temperature" way above that of water's boiling point even though the average temperature may still be well below the boiling point. If these "hot" molecules are near the exposed top surface of the water, then they can escape from the rest of the pack as a molecule of steam. Evaporation allows these molecules that possess a sufficient level of energy to shift from the liquid phase to the gaseous phase and escape if a convenient exposed surface is nearby.

Evaporation is only one form among many of observable quantum effects. Another example is what we were discussing about vacuum fluctuations and virtual particles.

Scientists have observed through their instruments a phenomenon related to evaporation in so much as they both involve fluctuations of quantum energy. Whereas the quantum energy of a water molecule may take on a spectrum of kinetic energy values as it receives and gives away its energy to other water molecules and its energy may fluctuate either above or below the average kinetic energy level of the water molecules in the kettle, "empty" space itself has been observed to possess varying levels of energy that fluctuate around the average minimum energy of our universe. These fluctuations are called vacuum, or quantum, fluctuations.

As the energy of the vacuum fluctuates above a crucial threshold, virtual particles are observed to first enter into existence and then disappear as the energy level then fluctuates back to a lower level. If the fluctuation lasts long enough and is sufficiently energetic, these virtual particles are observed to "become" real particles. They become real particles in the full sense that the observer perceives the full-enchilada spectrum of perceptions that are associated with these particles, just like any other real particles.

In laboratory settings, scientists can help the fluctuation process along and facilitate the emergence of the virtual and real particles. The perceptions of these scientists results in the impression that the vacuum fluctuation energy at any point in space is infinite, since the scientists are not bumping up against any perceptible energy limitations. Scientists then extend this notion of infinite vacuum fluctuation energy to each and every point of space in a universe believed by them to be infinite in size. Pretty soon you've got a universe possessing an infinite amount of energy, and you already know how I feel about that silly idea.

So, our next step involves rectifying this quirky perception resulting from misinterpretation of perfectly good scientific measurements that each and every point in the universe seems to possess a bottomless pit of potential energy. It would appear that once again we might have painted ourselves into a corner, but like usual in this book, it's not at all the case.

This is yet another phenomenon that combines with the biases resulting from everyday experience that cloud a better interpretation which would avoid the emergence of an associated paradox. Please recall my earlier assertion that it appears very much to me like the plethora of paradoxes arising from today's science are related to one another. Overcome one and many, if not all, might fall like the proverbial row of dominoes.

Chapter 22

Direct versus Indirect Observations

For the starting point of this journey, I chose the double-slit experiment, in part because of the deep admiration and respect I feel for Dr. Feynman and his insights. There doubtless exist a large spectrum of paths one could take to overcome the paradoxes. Anyone who attempts this journey must choose their own specific starting point. Be prepared for a bumpy ride as you first deconstruct what you, or others, think they know and then replace it with new interpretations that remain true to harvested scientific data.

In any event, I've chosen to recount two seemingly disparate remarks from Dr. Feynman regarding the double-slit experiment and the insights its results might provide about our universe. The first remark involved Dr. Feynman's belief that all of quantum mechanics, and a great deal of understanding of our universe, could possibly be derived from understanding the results of the double-slit experiment. The other remark involved his Sum-Over-Paths model with which the double-slit experiment results could be duplicated by the photons taking every conceivable path through the universe as they emerged from the lamp and ultimately smacked into the projection screen so we could perform our measurements. How in the world can these two assertions possibly be rectified?

Simple. Here's how. The way we perceive, interpret, and categorize our perceptions of our world lead us to the faulty impression of discrete particles, of either the matter variety or the boson variety. Both Einstein and the quantum mechanics community understood that, at their heart, these phenomena consist of small amounts of energy, or quanta, which form stable configurations that we can observe and measure. Einstein showed conclusively that there is equivalence between energy and mass and that the product of mass times the square of the speed of light enables one to calculate the equivalent amount of energy that any particular amount of mass represents.

"Twentieth century, I'd like to introduce you to the fifth century BCE. There's this Parmenides guy I'd like you to meet."

Do you remember that Parmenides defined his "atom" as the basic, most fundamental unit of matter, which the void was incapable of dissecting any further? The void isn't comprised of a bunch of discrete morsels of emptiness that sum up into something big and meaningful. It is a single, contiguous component that happens to be the single biggest feature of our universe by an inconceivably wide margin. The void doesn't splinter into fragments. Parmenides' void is the underlying substrate of the entire universe, but it's easy to ignore, since we can't directly observe it.

This is where it gets very tricky, so let's try to stay together. I'll explain the next concept from a variety of perspectives. If the first explanation doesn't make sense, hopefully a subsequent explanation in the following sections will.

To take this next step, and it's a doozy, we need to review some of our glossary entries and fold in the interpretations of the data and perceptions we've accumulated thus far. In fact, it's probably a pretty good time to review the terms in our glossary anyway, so I've reproduced our not-yet-complete glossary here for convenience.

Here is our glossary so far:

◊ **Consciousness** – the act of synthesizing information about the internal self with information about the external universe

◊ **Life** – (a) the state of matter that meets all three of the following criteria: (1) possessing self-awareness of internal states, (2) capable of perception, and (3) incapable of existing in an indeterminate state with an uncollapsed probability field since the state of matter comprising the living entity is under continuous self-observation; (b) the state arising from a self-observable arrangement of matter resulting from self-awareness of internal states.

◊ **Matter** – determinate states in space potentially observable (measureable) locally by electromagnetism possibly in combination with the weak or strong forces, or potentially observable at a distance directly by electromagnetism in combination with gravity.

◊ **Non-life** – inanimate objects and/or phenomena that meet all three of the following criteria: (1) not possessing self-awareness of internal states, (2) incapable of perception, and (3) capable of existing in an indeterminate state with an uncollapsed probability field when not observed.

◊ **Observation** – the synthesis of information by a living entity that connects self-awareness with specific information it perceives about the internal self or the external universe.

◊ **Observer** – a living entity capable of perception relevant to a specific observation or set of observations; for example, an observation involving electromagnetic energy requires light perception in the correct portion of the spectrum by the observer.

◊ **Perception** – the collection of information by the self about the internal self (internal states) or about the external universe (external states) through the capabilities of its unique sensory apparatus.

◊ **Self** – a living entity capable of conceptualizing a boundary distinguishing that which it considers internal ("itself") from that which it considers external ("the world" around it).

◊ **Self-Awareness** – the sense of self unique to each self.

◊ **Space-Time** – the temporal-spatial framework unique to each observer based on the inertial reference state of the observer.

◊ **Void** – a single phenomenon comprised of the collective, composite, indeterminate probability field representing the full potential energy of our universe, for which there is no direct observational relationship to a particular observer or set of observers, and which is therefore undetectable and unobservable.

The term we really need to focus on is "space-time." It's the framework by which and in which we assemble our perceptions and combine them into an image of our whole world. Space-time is the same framework in which we build and store the contexts with which we view our world.

Einstein demonstrated conclusively that space-time is tied to the observer, that each observer uniquely perceives time at a rate that is consistent with their motion through space. At the same time space warps and bends in a way that is consistent with the observer's perceived motion through space and causes time to fluctuate in a way that ensures that the observer measures a fixed, rigid, constant speed of light. Please recall also that the only means available to the observer through which he can perceive perceptions is through the electromagnetic force. As the observer travels through space-time, each and every observation that he could possibly make and each and every measurement of the time and/or space through which he passes rely ENTIRELY on carriers of the electromagnetic force, the diminutive mathematical construct called the photon.

Here's what we have so far. Every observation that an observer can make relies exclusively on photons. The perceptions resulting from these photon-based observations are unique for each observer and, as a consequence, the perceptions of space-time are unique to each observer. These perceptions of space-time and their contexts are processed and stored in our minds. Since this conceptualization exists only in the mind of the observer, it is therefore not the sole and exclusive property of the universe.

As I have mentioned, all any observer can hope to achieve through his or her senses is the perception of perceptions, and this process results in a disarmingly deceptive view of our universe. We view the universe through the deceptive impression that space and time are fundamental, consistent, unchanging, rigid properties of our universe, and they are not. Einstein asserted it and subsequent scientific measurements confirmed it. Space-time is somehow a property shared between the observer and the universe.

On the other hand, the speed of light is something upon which we depend as representing a fundamental property of our universe, and any observation of the speed of light relies exclusively on our diminutive photon friends.

Are you with me so far? So far, so good; let's forge ahead.

Chapter 23

Meet the Hemiverses

As we've discussed, the results of our perceptions, logic, and interpretations provide us with the understanding that the universe can be split into two domains: the void, which is directly unobservable, and the objects and phenomena that we, as observers, can directly perceive and measure through electromagnetism. Let's reframe this assertion in a slightly different set of ways. The universe can be split into two domains: the void and the observable, the latter of which we are card-carrying citizens.

So, let's name these two domains the *voidverse* and the *observerse*. Each observer perceives the universe in a unique way, meaning that space-time is unique for each observer. So we're ready for the final reframing of the earlier assertion. The universe can be split into two domains or what I call "hemiverses": the domain of the void (*voidverse*) and the domain of the observer (observerse), the latter of which includes all directly observable phenomena and objects and to which all perceptions of space and time uniquely derive.

All we've done so far is to reframe our perceptions and interpretations incrementally through a series of substitutions using equivalent concepts and statements. As I mentioned earlier, substitution is a method often used in math and science so we can convert a problem we don't like into something with which we can better deal.

Let's return to our two hemiverses, the voidverse and the observerse, and analyze where we stand on each.

There's the domain of the observer and this domain consists of the observer and everything that the observer can directly observe. These things include everything we perceive during our daily lives, everything we can directly detect and measure via electromagnetism, other observers, our internal universe, the fuzzy boundary separating each object or phenomena from the others and the fuzzy boundary separating our internal universe from our external universe, among a host of other things. This domain also includes our minds and every perception of every perception our minds have ever or will ever experience. Our minds house our concepts and perceptions of time and space in a "Euclidean" way.

Anything within the hemiverse known as the observerse is precluded from directly observing anything other than what resides in this domain, and observation in this domain relies exclusively on electromagnetism. Anything in this domain is precluded from directly observing anything outside of this domain.

Anything in this observerse domain is precluded from directly observing anything within the domain of the void; the domain of the void is directly unobservable to us. We're able to indirectly observe the domain of the void in so much as a puzzler can indirectly observe the absence of the final puzzle piece that appears to him or her only as a bounded hole in the picture of the puzzle. We're not seeing the world in its entirety, and as a result we don't see the whole world as it is. Again, we don't see the world as it is, we see it as we are and as we think. Euclidean concepts of space and time apply to the observerse, but not to the voidverse.

Now, let's shift our identity back to that of the anthropomorphized photon we discussed earlier. We're now taking on the perspective of the photon and perceiving the universe through the equivalent of its really tiny and really imaginary ears and eyes. We can perform this thought experiment in our minds, but since we reside only in the domain of the observer, it's no easy feat and we can achieve it only up to a certain point.

Recall also that when scientists apply their concepts and equations of space-time to a photon, they determine that to a photon the universe would appear as a point in which time and spatial dimensionality do not apply. From its perspective as an indeterminate probability wave in the voidverse, the observerse is unobservable to it. Observerse space-time exists exclusively within the observerse. Whatever may exist in the voidverse simply does not include observerse space-time. What then might the domain of the void "look" like?

The voidverse consists of the total potential energy of the universe, which is quite considerable but finite. The boundary separating the voidverse from the observerse would appear to have some interesting properties. We can discuss the properties of this boundary but we will never be able to peer beyond it, through any direct observational scheme. In fact, we can never hope to adequately conceive an image of what the voidverse looks like for two very good reasons. One, it's not directly observable to us, and two, the senses we possess are useless in a temporal-spatial environment different than the one in our observerse. Our senses simply can't and won't function. They weren't designed to. Our perceptions would be null and the perception of our perceptions would be null. That's that. Without perceptions, we simply have no framework with which to interpret and ultimately understand the world of the voidverse. The voidverse could only appear to us as the utter absence of perception, the ultimate nothing of nothings. "Being is, but nothing is not." Parmenides was at the top of his game all along.

Let's return to the boundary between these domains. This boundary has the unique distinction of being the one and only boundary that's left for us to dissect. No matter what the voidverse is like, the boundary would appear "real" from the perspective of either hemiverse. Observerse space-time applies on one side (here) and voidverse space-time applies on the other side (there). If the void verse does not possess our space-time, there's something different with respect to dimensionality on the other side of the boundary. We have observerse space and time, and they don't. Let me say this again, because it's profoundly important. The observerse possesses the familiar dimensionality of space and time, and the void domain does not. This little tidbit has sweeping consequences, let me tell you.

So, let's talk more about this boundary. Energy crosses the boundary all the time in each direction as objects and phenomena switch between their indeterminate unobservable states and their corresponding determinate, observable "condensate" states.

Just like the books that librarians place back on shelves when they're returned by readers and that readers remove from same shelves to bring home and read, energy shifts relentlessly back and forth in a never-ending game of ping-pong as potential energy in the form of unobservable indeterminate probability waves collapse upon observation to measureable, detectable values. The total energy belonging to the books of the library never changes, but just as books move out and then back in the library, energy flows back and forth between potential and kinetic energy as the books are moved about our universe, or, perhaps more accurately, the observerse.

And, in keeping with the book analogy, the change in kinetic energy is rather easy for us to observe as we grab the book from the shelf, place it in our arms, check it out for loan at the front desk, and drive it to and from our homes. As we sit in our comfy reading chairs, repeatedly raising and lowering the book, our actions change the balance between potential and kinetic energy, but never do our actions alter the sum total of energy possessed by the book.

When a book is returned to the library and placed back on its shelf, it has returned to the energy state that it had previously occupied with the particular balance of potential and kinetic energy it once possessed. No one would even know that the energy state had otherwise changed and returned. In fact, other than wear and tear, there would be no way for an observer to detect it unless they observed the book during energy transitions. We notice these changes through their observable effects, and the only side of the energy equation we can directly observe is the change in the kinetic energy portion.

As many of us might recall from high school physics, the very concept of potential energy seemed completely baffling at first because there's simply nothing to observe when an object is just sitting there minding its own business. The concept of the void represents the very same sort of experience and attempts at its conceptualization have baffled scientists and philosophers for thousands of years.

Our observations, especially at the quantum scale, are comprised almost entirely of detecting and measuring the transitions that occur between one hemiverse and the other. In fact, a couple of good examples of this transition are called wave-particle duality, the speed of light, and entanglement. I'll finish exposing these phenomena for what they are and eliminating the supposed mysteries enshrouding their associated paradoxes once we tackle our one, remaining enigma and complete our glossary. Our next order of business is to tackle the phenomenon called "mass."

Chapter 24

The Matter of Mass

We've come to consider mass as a basic component of our existence. The concept of mass, and its close cousin, weight, pervade all aspects of our daily lives. Our physicians measure and record its value in our medical records as a key indicator of health and well-being. We purchase our nourishment in the form of food based very often upon its mass and we gauge its nutritional value by the mass value of its ingredients. The gravity arising from an object's mass ensures that we stay firmly on the ground of Earth, that the moon and Earth continue in their orbits around one another, and ensures that, together, the Earth and moon remain in orbit around our sun, the source of energy that sustains life on our amazing planet.

Yet mass and its related effects, involving gravity, remain among the least understood and most elusive concepts of modern science. At the moment, science has absolutely no real clue about the nature of mass and momentum, which is in essence the energy possessed by a kinetically moving mass that is preserved in any closed system. (Momentum is the multiplication product of mass and velocity.) An incomplete understanding of mass and momentum, combined with incomplete understandings of time and space, likely are the elements that make Zeno's paradoxes against motion so mysterious and deliciously inexplicable. Zeno's paradoxes have survived with nothing more than small nicks, dents, and scratches for thousands of years.

Zeno of Elea (ca. 490-430 BCE) was a student of Parmenides and a member of his Eleatic School. Although little is known today about Zeno, some of the paradoxes he illustrated have survived. In fact, Aristotle included some of Zeno's "paradoxes against motion" in his own book, elegantly and simply named *The Physics*.

The drive to understand mass has consumed a great deal of the attention, energy, and funding of modern science. In fact, the craving to understand the nature of mass has to a large extent driven the design and construction of ever-more-potent particle colliders. At the root of this drive is the mathematically hypothesized Higgs boson. Many theoretical and mathematical scientists believe that the Higgs boson is the carrier of mass and further believe that its discovery may be right around the "particle accelerator corner." The Higgs boson is theorized to possess an energy equivalent that may even be within the range achievable through the Large Hadron Collider (LHC). The range of possible energy values hypothesized for the Higgs boson, however, extends up to a value that is far beyond the capabilities of the LHC. So researchers have their fingers crossed that the observation and discovery of the Higgs boson will be within the incredible, but limited, capabilities of the über-expensive collider at CERN. The rough total cost of the LHC? Nine billion dollars. The value of the experiments performed using the LHC? Priceless, or at least that's the hope of the scientists who devised it and the governments that paid for it.

So, we've come upon another complex mathematical construct of modern science. For all I know, the mathematical construct of the Higgs boson might be right on the money, and believe you me, I'll be in the stands waving my pennant and cheering like the lunatic parent of a grade-school soccer player with the rest of the science community if it comes to be. Personally, I want the Higgs boson discovered during my lifetime, so I can add it to the list of "Amazing Events that Actually Occurred During My Lifetime" like the Apollo 11 moon landing, the establishment of the World Wide Web, the sequencing of the human genome, and the "discovery" of dark matter and energy. Many people may construct a "Bucket List," but I think that science-types like me are generally more concerned about experiencing as many important scientific achievements as possible before we depart this existence.

In any event, mathematical constructs exist only in our minds. The particular construct called the Higgs boson may or may not have a basis in reality. Whether or not it does, the mathematical model of the Higgs boson is an attempt at a representation of the currently mythical Higgs boson. Taken from that viewpoint, this model, like other models, is an attempt to capture and comprehend a partial, simplified understanding of some aspect of the universe. As we have discovered on the journey in this book, these non-humanistic mathematical models may capture key elements of a concept and perhaps even do so with great fidelity, but the nature of human perception and understanding also requires a personal context with which to make the concept more meaningful and embraceable.

And so we've arrived at the point in our journey at which, once again, we review key elements that we have previously discussed and considered so that we can build these elements up into new, meaningful interpretations and contexts.

Here's what we've got so far. Through observations, we know of four fundamental forces: the strong force, the weak force, electromagnetic force, and gravity. The relationships among most of these forces are fairly well understood, and science has made, and continues to make, great strides in "unifying" the strong, weak, and electromagnetic forces. These three forces are tightly coupled to fundamental particles of matter as described within the Standard Model. The strong force, or strong nuclear force, is credited with keeping protons together in the nucleus and overcoming the repulsive electromagnetic force arising from the proton's "+1" electrical charge state. The protons themselves are each comprised of quarks held together by gluons. The weak force, or weak nuclear force, is credited with keeping neutrons stable and bound within the nucleus despite the absence of an electrical charge state. (The neutron's electrical charge state is "0.")

The electromagnetic force is carried by photons that exist in an indeterminate form until an observation is performed, forcing the photon into a determinate, and therefore observable, state whose properties we can then measure. Each of the particles associated with these three forces and the associated carrier particles are comprised of energy, which as we know from Einstein's Special Theory of Relativity possess an equivalent amount of mass into which the energy may be converted (according to $E = mc^2$). The rest mass of the force carriers is zero, but each of the particles of matter, or their antimatter counterparts, possesses attractive gravity. No "negative" gravity is observed, directly or indirectly—only positive, attractive gravity. (Let's again ignore dark energy for now.)

Furthermore, Einstein's General Theory of Relativity tells us that the basis for gravity is actually the curvature of space resulting from the mass possessed by matter. As a result of mass, space becomes curved inward toward the mass' center of gravity.

Special relativity also tells us that the force arising from gravity is equivalent to the force resulting from acceleration. In other words, general relativity asserts that gravity and acceleration each arise from the curvature of space resulting in one way or another from the presence of mass (or energy). Gravity or acceleration is described as a change in velocity per unit time. In the case of distance measured in meters (m) and time measured in seconds (s), the units are m/s^2. Einstein accurately described the curvature of space due to the presence of mass in exquisite detail but did not attempt to explain the basis underlying space's inward curvature.

From our attempts to understand the void (or the voidverse), we developed an understanding that the dimensions of observerse space-time are different in some way from the dimensions of the voidverse. We also developed an understanding that the voidverse does not or may not possess the familiar dimensions of observerse space-time. Nonetheless, the void, which remains otherwise directly unobservable from the *observerse*, manifests itself as empty, or "free" space in the observer domain through which observers can and do travel with apparent ease.

We also learned that for a typical atom (we had chosen carbon), less than one part volumetrically in 500 trillion parts is comprised of matter while the rest is comprised of free space, a.k.a. void. So, if we calculate and compare the mass density of diamond (a particular crystalline form of carbon) against that of our own galaxy, the Milky Way, we end up with comparative volumetric matter densities of 1 part of matter per more than 500 trillion parts of empty space for diamond versus 1 part of matter per more than 10^{23} quadrillion parts of empty space, indicating that the difference between the hardest material known to man and the sparse distribution of mass-energy in our galaxy is the difference between two exceedingly small values. (See Figure 9.)

	Observed Mass Density (approximate)	Calculated Volumetric Matter Density (ratio of matter volume to empty space volume)
Diamond	3.5 g/cm^3	< 1 : 10^{15}
Planet Earth	5.5 g/cm^3	< 1 : 10^{15}
Milky Way Galaxy	1.9e^{-23} g/cm^3	< 1 : 10^{38}
The Universe	1.9e^{-26} g/cm^3	< 1 : 10^{41}

Figure 9. Mass Density versus Matter Density.

We can feel confident in our assertion that empty space, or void, occupies space in our observer domain, because we can perceive that bits of matter are held apart from one another by the presence of empty space, which we can routinely and consistently measure using units of observerse distance and time. From our discussion, we also understand the prospect that potential energy stored in the voidverse as indeterminate probability waves, cross the boundary into the observer domain where the probability waves collapse to determinate states we measure as carrier bosons (like photons) or as particles of matter (like electrons and quarks).

There is nothing that we have yet detected in our observations of particles of matter or particles of energy that provide us with insight into the property we call mass, other than a hypothetical mathematical construct that scientists refer to as the Higgs boson.

Finally, there's an aspect of gravity that many of us might already take for granted. We are incapable of measuring gravity directly. Our perceptions of gravity derive from indirect observations involving electromagnetism. Another way of wording this is that we depend directly on observations involving the behavior of luminous matter and light (photons) to indirectly observe gravity. Hmmm, this sounds familiar, doesn't it?

From where I'm sitting (and I've been sitting and thinking about this for quite some time now) there's only one way to combine all of this stuff into a single, self-consistent picture. So it's time for my next hypothesis.

Mass is not a property of matter; it is a property of space.

More accurately, mass is a property of the voidverse manifesting itself in observerse space-time. Let's take another of our backward steps and break down this new perspective.

According to my hypotheses, the voidverse is simply the "place" where the universe stores its enormous supply of potential energy, while the observerse is the place where "kinetic" objects and phenomena interact with each other and observers to create the perception of our universe that's the starting point of this whole discussion. I refer to the kinetic manifestation of objects and phenomena as "condensates" possessing real, actual, determinate energy not of the "potential" variety.

The universe is a closed system possessing an unimaginably high, but finite, total energy. Energy can, and does, routinely flow in both directions back and forth between the voidverse and the observerse. Observers within one domain are incapable of direct observation of the other domain because the dimensionality of the two domains differs from one another in some way, which we'll discuss in greater detail later. Observerse space-time exists solely in the observerse where it is observed and perceived to be unique to an individual observer based on aspects of their motion through empty space. These perceptions arise from direct observations of electromagnetic phenomena and from indirect observations of mass and gravity via electromagnetic phenomena.

In fact, direct observations in the observer domain are limited to electromagnetic phenomena. Observers in the observer domain are precluded from performing measurements and from acquiring direct perceptions of the voidverse beyond the boundary separating the two hemiverses, since our direct observations are limited to electromagnetic phenomena occurring within observerse space-time, which exists only in the observerse.

Potential energy in the voidverse is therefore directly inaccessible to observation by observers in the observerse until energy shifts to determinate states that observers can directly detect and measure. Science has developed mathematical tools and disciplines, such as quantum mechanics, to calculate the probability field in the observerse associated with the indeterminate state of potential energy in the voidverse. Each object or phenomenon we observe results from the collapse of the probability wave function described by quantum mechanics associated with that object's or phenomenon's contribution to the probability field. The full potential energy possessed by the voidverse manifests itself in the observerse as free, "empty" space, but this free space is formed in our domain by the composite indeterminate probability field. In other words, the entirety of the indeterminate probability field is equivalent to the voidverse and manifests itself in the observerse as the empty space separating the discrete objects and the phenomena we observe, perceive, and measure.

I know this is a lot to chew on, but allow me to continue.

As a probability wave collapses to an observable, determinate state, a small amount of energy (a quantum) is transferred from the voidverse to the observerse. Since we perceive the potential energy of the void domain solely as empty space, then the transfer of potential energy out of the voidverse incrementally reduces the local potential energy of the voidverse. In the observerse, this change in potential energy of the void manifests itself as an incremental reduction in the volume of voidverse space. In essence, the transfer of energy from the voidverse to the observerse results in the collapse of the quantum's probability field to a determinate state that we perceive as an object (particle) that condensed from the quantum fluctuation energy. As the particle condensed into matter, the "empty" space that the probability wave previously occupied (potential energy in the voidverse) reduced in volume with the result that we perceive the volume that the probability wave previously occupied in the observerse shrinking in extent, or curving, which we can observe only indirectly via electromagnetism.

This is probably more information than you might have cared for, so let me boil it way down.

Mass results from the shrinkage of space associated with the transfer of potential energy from the voidverse to an object or phenomenon that can be directly observed in the observerse. The shrinkage of space resulting from this transfer is what we call "mass." Mass is the curvature of space resulting from this shrinkage. The transfer of potential energy results in the associated and necessary observation of matter at the heart of this puckered, inward curving space. We refer to "gravity" as the effect that we indirectly observe via electromagnetism of the interaction between two masses.

This seems like a good time for me to go one step further and add to this hypothesis that anything that lowers the potential energy of the void creates gravity in the observerse. We'll talk more about the implications of this in the next section that discusses the gravity-acceleration equivalence of Einstein's General Theory of Relativity.

Chapter 25

A Short History of My Take on

Science and Philosophy

Let's summarize what we know and the hypotheses I've proposed before we begin to discuss their implications in the sections that follow.

First, people started detecting paradoxes about "reality" shortly after the emergence of homo sapiens, once they had taken care of the basic needs of food, beverages, shelter, and clothing. During the sixth century BCE, Greek society had risen to the point where life was enjoyed a bit more leisurely and people could spend more time thinking about the world and discussing it amongst themselves. During this period, Parmenides, along with the other pre-Socratic Atomists, developed certain key insights and hypotheses about the nature of reality, from which the scientific method would ultimately emerge. Among the crowning achievements of the Atomists was the hypothesis of tiny, fundamental bits of matter that could not be divided further by the void. Concepts of the void, however, also divided the Atomists into two groups based on whether or not the void "exists" or represents the quintessential "non-existence."

In line with these insights and hypotheses, Zeno's keen sense of observation enabled him to eloquently verbalize paradoxes with which humanity has been grappling ever since. We're now in the third millennium following Zeno's death and these paradoxes remain unresolved and continue to capture the imagination of those who encounter them. Among the key concepts introduced by the early Greek philosophers was the notion that our senses could not be counted upon exclusively to provide us with a true interpretation of reality, and that abstract reasoning was required, in whole or in part, in order to create a complete, self-consistent interpretation of the universe and the phenomena we call reality.

Although not formulated mathematically until the end of the nineteenth century, the concepts of the conservation of mass-energy and those of relativity were first introduced by Greek and Eastern philosophers between the sixth century BCE and the second century CE, while the concept of the void continued to be discussed and refined. Nagarjuna provided some key insights about the natures of existence and reality during his work to extend and comprehend the four states of being, known as the *Tetralemma*, earlier devised by Greek philosophers. Nagarjuna also understood the dangers that doctrinal thinking represented in attempting to understand the complex universe in which we live.

Greek philosophical thinking culminated in Aristotle's concept of the physical universe and its subsequent establishment as the final word on science and physics for an astonishing 2,000 years, until Sir Isaac Newton introduced science's second final word on the physical understanding of our universe. The laws of physics introduced by Newton at the end of the seventeenth century provided a seemingly accurate portrayal of a broad spectrum of everyday experiences and scientific observations but, ultimately, paradoxes emerged and its limitations became apparent to scientists. Newton got somewhat of a raw deal because his physics stood for less than 300 years before it was abandoned by science as flawed and incomplete, but the pace of science and progress was picking up steam.

The first attempts to measure gravity were performed by Henry Cavendish at the very end of the eighteenth century and his experiments provided the first estimates of the value of the universal constant of gravitation incorporated within Newton's laws. Scientists have continued to refine their estimate of its value ever since.

Before the end of the nineteenth century, Maxwell and Lorentz had developed concepts and mathematical schemes for better understanding the nature of electromagnetism and the nature of observations based upon information carried by electromagnetism. The science community was both divided and baffled about the way in which electromagnetism traveled through space, so theories like that of the Luminiferous Aether emerged and abounded to offer explanations seemingly consistent with scientific observation.

Einstein, a lowly patent clerk in Munich, integrated the concepts into a coherent, internally consistent scheme that built upon the work of these and other science giants. This led him to develop and publish his Special Theory of Relativity in 1905, science's third final word on the physical understanding of our universe. It was subsequently and firmly embraced and established. This Special Theory of Relativity showed conclusively that the speed of light for any and all observers is the same. Theories involving the Aether were abandoned but, curiously, a complete physical understanding for how electromagnetism propagated through empty space was never developed. In fact, this concern, along with the concept of the void, was largely but not entirely swept under the cosmic rug.

In 1915 Einstein published his General Theory of Relativity, which provided a detailed mathematical description of gravity that also demonstrated acceleration and gravity were equivalent to one another—science's way of saying that they are the same thing. The General Theory of Relativity became science's third final word on the physical understanding of our universe, Part B. Together, the combined theories of relativity provided some further insight into cosmic gravitational observations while also providing unprecedented insight into the workings of the universe, at least on the large scale.

Continuing attempts by mathematical scientists to calculate the universal gravitational constant from first principles continued to fail miserably, at best.

Also during the late nineteenth and early twentieth centuries, other scientists began to develop an understanding of objects and phenomena observed at the other, really small end of the observational spectrum. The model that developed is called quantum mechanics, since it deals with the statistics of exceedingly small packets of energy, called quanta. Quantum mechanics, and its newer variants, quantum chromodynamics and quantum electrodynamics, have been wildly successful in describing the behavior of particles and most of the known fundamental forces (with the exception of gravity) at the very smallest of scales. Quantum mechanics vies for the enviable distinction of being science's co-third final word on the physical understanding of our universe, along with relativity.

By the middle of the twentieth century, it became clear that the two co-third final words on the physical understanding of our universe, quantum mechanics and relativity, were incompatible with one another and provided inconsistent results when applied to the domain of the other's theory. During this time, a peculiar phenomenon called entanglement was observed and met with mixed reviews by the science community. Einstein refuted the notion of entanglement as "spooky action at a distance." That introduced a bit of tension between him and the quantum mechanics community. He published a paper with colleagues on what would become known as the EPR Paradox, and this became the most highly referenced paper in all of modern science. Simultaneously, the Copenhagen Interpretation of Quantum Mechanics asserted that there's no reality other than what can be observed and measured.

The 1930s brought us some of the most creative interpretations of scientific observations. An ordained priest and professor of physics, Georges Lemaître, proposed his "hypothesis of the primeval atom," and this became the basis for what evolved into the Big Bang Theory. This theory achieved and maintained celebrity status despite the fact that it introduced more paradoxes than perhaps anything else in the history of science.

Also during this time, an astronomer named Edwin Hubble observed a shift in the spectrum of light reaching us from distant luminous celestial bodies. With Einstein's acquiescence that perhaps he was wrong about the cosmological constant in his relativity equations, he concluded that this was ample evidence to support the interpretation that the universe was expanding—and, oh, by the way, everything in the universe was moving away from us. But that's fine because that must also mean that everything was also moving away from everything else. So, everything was fine even though the edges of the universe were moving outward at a rate faster than the speed of light. Mathematical physicists made fantastic progress in creating stunning, new ways to make the theory of the Big Bang better fit what scientists actually observed. Paradoxes continued to abound and science progressed at an ever-quickening pace.

(For getting Einstein to admit he was wrong about anything, whether or not he actually was, humanity showed its appreciation to Hubble by naming a telescope after him that would be placed into operation for a few years later in the century and into the start of the next. Georges Lemaître, the originator of the Big Bang Theory, on the other hand, faded into relative obscurity, perhaps in part due to the fact that he hadn't gotten a single admission of error from Einstein.)

Continuing attempts by mathematical physicists to calculate the universal gravitational constant from first principles continued to fail miserably.

In the deepening struggle to overcome the growing number of paradoxes emerging from the divided house of science regarding the continuing inconsistencies between relativity and quantum mechanics, a slew of fourth final-word-on-the-physical-understanding-of-our-universe wannabe theories emerged, got us all really excited by the prospects they offered, then stagnated, then re-emerged with new twists, then just sort of lingered, or perhaps hovered, making everyone feel a little unsettled. Among these theories was String Theory and its variants, each of which say pretty much the same thing but that provide no basis by which scientific methods can be used to evaluate them.

At the end of the twentieth century, space-based observations and experiments revealed that the universe was a peculiar place that we didn't know all that well. The observations indicated that the cosmic infrared and the cosmic microwave backgrounds were really, super uniform, and no matter where we looked, it seemed like we were at the center of a rapidly expanding universe. (Hmmm, haven't I heard something about us being in the center of the universe before?)

At the beginning of the twenty-first century, in deep contrast to the century that preceded this particular century by one year, new, technically advanced, and immensely more expensive space-based observations and experiments revealed that the universe was still a peculiar place. The observations indicated that the universe was fraught with paradoxes and that the cosmic infrared and cosmic microwave backgrounds were really super uniform; the results indicated that no matter where we looked, it seemed like we were at the center of a rapidly expanding universe. That, and the fact that we missed 96 percent of the mass-energy of the universe in all of our previous observations. From these new observations, scientists concluded that the 4 percent of the mass-energy we're used to seeing was still there—plus there was another 20 percent they called "dark matter," and another 76 percent they called "dark energy"—because these terms are really mysterious. That, they thought, ought to distract everyone from noticing that we had absolutely no idea what dark matter or dark energy might be. Through these observations and the resulting analysis, science once again confirmed that the sum total of the mass-energy of the universe remained fixed at 100 percent. Science definitively pinned down the age of the universe at 13.7 billion years.

Candidates jockeying for the position of "fourth final word on the physical understanding of our universe" proliferated at an almost inconceivable rate, as the latest of the really important scientific discoveries prepared to enter middle age and the world's most costly science experiment, the Large Hadron Collider, began operation in the hopes of giving science some clue about where to turn next.

Continuing attempts by mathematical physicists to calculate the universal gravitational constant from first principles continued to fail. The ongoing befuddlement prompted gravitational physicists to raise the volume on discussions involving gravitational waves and the carriers of the gravitational force called "gravitons" to begin paving the way for space-based gravitational wave experiments the size of our solar system, since they began to run out of schemes to consistently fail to observe gravitational waves in terrestrial settings.

In 2010 I read Dr. Lanza's *Biocentrism* with the intent to denigrate it since Dr. Lanza is a biologist and not a "hard" scientist. Instead, I discovered to my deep surprise that *Biocentrism* provided me with some key pieces I've been missing from my own attempts to construct a cohesive, internally consistent understanding of the universe.

In my quest to figure out the meaning of everything, I made the conscious decision that one unknowable universe is more than sufficient, thank you, so I'll just pass on all the colorful variants involving a plethora of unknowable universes. Instead, I proposed a model based on the notion that the universe can be divided into two segments in yet a new way that involves two hemiverses. Failing to actually suggest anything to which science was not already deeply accustomed, I proposed that one hemiverse contains the potential energy of our universe and the other hemiverse contains the equivalent of the universe's kinetic energy, which is really quite dull when you realize that it resembles some of the more boring content of high school physics that science considered quite rudimentary even back in the 1970s, when I was sitting in class. Together, the two hemiverses possess the sum total of the universe's energy and the universe's energy can, and does, move freely across the boundary between these hemiverses in both directions as the balance of energy shifts between kinetic and potential energy, very much like an old-fashioned spring common to our everyday experiences, as well as 1970s-era physics.

In one of these hemiverses, the observer hemiverse or *observerse*, you'll find all the stuff we can, and do, regularly directly observe. This hemiverse includes all of the objects, phenomena, and observers with which science is keenly familiar and by which direct observation (Class 1 observations) is routinely achieved through electromagnetic-based schemes within an observer framework comprised of observerse space-time. This hemiverse also appears to contain a colossal amount of empty space upon which we are unable to perform any direct observation or measurement, but which we study and analyze using indirect methods of observation (Class 2 observations) based upon a combination of gravity-based and electromagnetism-based schemes. This nothingness appears to isolate each of the objects and phenomena we perceive in the space-time of our observerse. Also, the best place to find observers is in the observerse.

The other hemiverse, called the *voidverse*, contains all of the potential energy of the universe. We can't view this voidverse because there's nothing that we could consider determinate in space or time or upon which we could successfully perform a measurement. We can't see beyond the barrier separating the two hemiverses, but we sure can detect the consequences of energy shifting one way and the other across the boundary. As an observer performs an observation or takes a measurement, the energy flow takes the form of probability waves collapsing into objects that we perceive as possessing determinate states. Following its transformation from potential energy, the particle of matter we observe possesses an amount of "kinetic" energy equivalent in magnitude to the amount of potential energy from which it formed, which can be calculated using the equation $E = mc^2$. If the transformation from potential energy to "kinetic" energy, which takes the form of directly observable objects and phenomena, is incomplete, we perceive the creation of virtual particles from the void.

As the energy is transferred from the voidverse to the observerse, the amount of potential energy is incrementally reduced. Since we perceive the universe's potential energy as empty space due to our inability to directly observe it, the reduction in potential energy manifests itself to us as a puckering of void space as the former state of its empty space shrinks into its indirectly observed, new, collapsed volume state with a morsel of matter at its center. The curvature of space resulting from this puckering provides the morsel of matter with the characteristic we perceive as mass, and the resulting interactions between all of the morsels of matter we observe in the universe is the phenomenon we refer to as gravity. Mass is not a property of matter; it's a property of space.

The implications of my proposition and its hypotheses are staggering, and my proposition provides a scheme by which the paradoxes of modern science and ancient philosophy may finally be eliminated once and for all, that is at least until the fifth "final word on the physical understanding of our universe" emerges.

If all of this isn't completely clear to you yet, please continue on to the next section of the book where I discuss key implications of my theory and attempt to provide a greater context by which to understand it.

Section III

Implications, Confirmations, and Demonstrations: The End of Paradox

Chapter 26

Zeno Enters Retirement

In section III, we'll build up our understanding by applying the thoughts pre-sented earlier in the book to paradoxes old and new. These exercises will give us the opportunity to put my theories into different contexts and to extend them further. These discussions will also give us the opportunity to discuss the relation-ships between my theories and the theories of relativity. Hopefully, my theories will become easier to understand and embrace as we broaden our perspective through these discussions.

The paradoxes that Zeno described are based upon common, familiar, every-day experiences and seem to have defied every attempt to use logic as a means of overcoming them. For nearly 2,500 years, Zeno's paradoxes continue to baffle the attempts by scientists, philosophers, mathematicians, and logicians, as well as anyone else who has attempted to contemplate and overcome them. Any attempt to rationally and logically approach the paradoxes seems to result in digging a deeper hole from which to emerge.

I first encountered Zeno's paradoxes in high school math and philosophy classes, so they've been stewing around in my head for quite some time, but my other thoughts about the nature of the universe pushed Zeno and his paradoxes into a deeper recess of my mind. I decided to examine the paradoxes again in the context of the theory I proposed. I found, quite interestingly to me at least, that my theory might provide a framework by which to overcome them.

In particular, I think the "moving arrow" paradox against motion is par-ticularly well-suited to attack. In this paradox, Zeno describes an arrow that has been let loose by an archer and is traveling toward its target. As the arrow travels through time and space, it will pass through each of the points in space corresponding to the trajectory it follows.

Let's divide up the situation into a whole lot of "snapshots" that represent the arrow and its position at various instances of time. For the sake of any photography buffs reading this book, we'll take our imaginary snapshots using our fastest shutter speed, a camera setting that represents the "action" or "sports" setting commonly found on digital cameras. This will ensure that each picture shows the arrow in a crisp, clear fashion. Now, let's examine a picture taken from somewhere in the middle of its flight. For that instant of time, the arrow appears to be standing still or, more accurately, it looks like it is stationary with respect to its background; it appears suspended and motionless in the air.

If we view this snapshot and think about it, the question arises: "If at any vanishingly small 'instant' of time the arrow can accurately be portrayed as 'standing still,' how does the arrow move from the position it occupied in space beforehand to the position in which I now observe the arrow?"

Then, I have another question: "If the arrow is stationary at some point in time, then how does it continue to move into the position in which I will next observe it?" Of course, the same question needs to be asked at each point along the trajectory of the moving arrow, but there would be little point in continuing to incorporate the remaining questions in the series, since you probably get the gist.

So, for nearly 2,500 years, after each person examines and ponders the paradox, only two possibilities arise for understanding the situation described by Zeno in the paradox. Either (a) motion is an illusion, or (b) there's something missing from our understanding of the arrow's motion. This type of paradox has led many a philosopher to the conclusion that motion is an illusion, but it's awfully hard to subscribe to that view when you're rubbing a recently stubbed toe.

Armed with both my new theory and numerous memories of stubbed toes, I decided that based upon my perceptions of perceptions, and the understanding that formed within my mind, that motion is not an illusion. If I observe motion, then there is some "real" phenomenon that I'm observing. As we discussed earlier, since I'm unable to observe something that flat out does not "exist" as an object or phenomenon in my world, I am observing something "real." Therefore, I choose option "b."

Having made that choice, I see a way to pummel the obnoxious little paradox that's been frustrating us for so long.

Let's review some of our earlier discussions about geometry first, though, before we proceed.

Geometry students are presented with some apparently conflicting pieces of information. First, the instructor informs them of the definition of a "line" in a flat, Euclidean plane, something along the lines of "a one-dimensional, infinitely long, and perfectly straight geometric structure." Sometime after that, the instructor teaches the students about Euclidean planes, and the students are provided with a definition along the lines of "a perfectly flat, two-dimensional, surface that extends infinitely."

At some point, the teacher inevitably combines the two geometrical concepts and introduces the concept of parallel lines, which they might define as something like, "two lines that lie on the same plane a uniform distance apart along their entire lengths." Fine so far. We accept the lines, we accept the plane, and we even accept the notion of infinity. Of course, the whole time, as students ourselves, we grappled with the "unrealistic" aspects of each of these concepts, since we don't ever encounter such ideal structures in the real world. Of course, some student will inevitably comment to the teacher that these things can't possibly exist in our world. The teacher calmly explains that these represent "ideal" structures that don't actually exist; we're just going to pretend that they exist. We choose to accept what he or she tells us unless we want a miserable grade.

Later in the class the teacher drops a bombshell on the students. The teacher may mention that the only place where parallel lines intersect or meet is at infinity. What? No way!

Let's think about what this innocent assertion means. We have two infinitely long lines (which don't really exist) on an infinitely wide plane (which doesn't actually exist). These two lines are uniformly spaced from each other along their lengths (hmmm, I guess that's okay, if only the lines and plane actually existed). Now extend the lines that don't exist on the plane that doesn't exist and the "parallel" lines meet only at infinity, which also doesn't exist.

Nonetheless, there's something terribly revealing about even verbalizing the notion that "parallel lines meet at infinity."

Place yourself, once again, in the center of a really long stretch of highway. (Disclaimer for those playing along at home: Please do not attempt this exercise anywhere else but in your mind! We're just going to construct this image from our memories, okay?) As you look down the length of the highway (in your mind!), it appears to converge in the far distance to a point. Artists capture this element in their drawings as "perspective."

So, the very concept of parallel lines converging or intersecting at infinity really combines two very different elements: the ideal notion of lines and planes in our minds, and our observations of the "real" world. The concept of parallel lines meeting at infinity is incongruous from either perspective individually.

Now, let's return to Zeno's "flying arrow" paradox and focus on our understanding of the arrow's motion. We're going to break down the problem as we normally do into its elements, and then reconstruct a new understanding.

Based on our understanding of physics, we know that the moving arrow possesses something called momentum. Momentum is calculated by taking the multiplication product of an object's mass and velocity.

What is our basis for determining the object's velocity? We directly observe the velocity of the object by measuring its rate of motion; this is done by dividing the distance travelled by the object by the time over which the object travelled. After all, as observers, we own time and space, so determining a velocity is something we can do quite well through direct electromagnetic-based perceptions.

What is our basis for determining the object's mass? Well, according to my theory, any measurement of an object's mass is an indirect measurement involving the reduction of potential energy into directly observable particles that we call matter. The mass of the arrow is simply a quantification of the total curvature of space resulting from the collapse of the indeterminate probability waves in the form of potential energy residing in the void hemiverse. Mass is a property of space and not a property of matter.

As a consequence, similar to the notion of "parallel lines meeting at infinity," the concept of momentum possesses an internal incongruity that itself can only be noticed from the perspective of our new understanding. The paradox arises as a result of mixing and matching phenomena, which at their root are fundamentally different. The physical basis for one phenomenon (motion) is entirely different than the physical basis of the other phenomenon (mass). One phenomenon is based in the observer hemiverse and the other is based in the void hemiverse.

I'm in no way implying that the concept of momentum is a flawed, unworkable concept. It's nothing of the sort. In fact, in so many ways, the concept of momentum, in one form or fashion, is at the heart of each and every attempt to construct a meaningful understanding of our universe. I'm saying that the reason why the concept of momentum is so important is that at its very heart it bridges the two hemiverses of our universe and captures the essence of our universe in a meaningful way.

I'm also just saying that if mass is not what we thought it was, neither is momentum.

So let's go back to our snapshots of the arrow in motion. We can take these snapshots and observe the arrow as a series of stationary objects at each instant of time during the arrow's flight. This part of our observation is a direct observation of matter in the framework of space and time in our observer hemiverse. On the other hand, the momentum that carries the arrow forward from one frame of time (snapshot) to the next involves two components: mass and velocity. The arrow's mass is an indirect observation of the manifestation of a phenomenon from the void hemiverse. The momentum that serves as the basis for carrying the arrow forward to its next position is, in part, comprised of a component that could never be captured by our snapshots. The paradox only arises as a consequence of the picture being incomplete. The paradox arises from the failure to make a distinction between Class 1 and Class 2 observations.

There's absolutely nothing wrong with the universe. The paradox arises from our faulty and incomplete understanding of mass and the nature of our universe. The paradox arises because we're combining two ingredients from two entirely different pantries. The paradox begins with a faulty, incomplete snapshot depiction of the arrow's motion, which captures the directly observed space-time elements (position and velocity) but completely neglects to capture the directly unobservable (or indirectly observable) element of mass. One phenomenon arises from the observer hemiverse, and the other from the void hemiverse. It's like separately building the ends of a bridge, one end in one hemiverse and the other end in the other hemiverse, and wondering why the bridge doesn't meet in the middle.

Neither individuals nor physics has previously made any distinction of its laws and concepts on the basis of which phenomenon arises from which hemiverse. Without this distinction, paradoxes abound. The universe does not create paradoxes, after all, but a faulty and incomplete doctrine does.

Let's take this thinking one step further. Let's get rid of the void hemiverse entirely for the moment. (Once again, if you are playing along at home, do this only in your mind, since doing it in "reality" would have disastrous consequences!) If there were no void hemiverse, there would be no potential energy. If there were no potential energy, there would be no empty space for us to perceive. If there were no empty space, motion itself would not be possible. If motion were not possible, there would be no such thing as momentum. Of course, if this were the case, there would also be no such thing as observers, so the point is perhaps somewhat academic.

A similar scheme can be applied to Zeno's other paradoxes, including the "dichotomy class" of paradoxes like "Achilles and the Tortoise." These paradoxes also involve motion through space. Movement through space can no longer be viewed as either (a) crossing an infinite number of spatial points in a finite amount of time or (b) the computation of a mathematical series comprised of an infinite number of terms that need to be summed in a finite amount of time.

Chapter 27

Implications Involving Feynman's Sum-Over-Paths Model

L et's return to the double-slit experiment and review now what Dr. Feynman had to say about it. Recall that Feynman believed that all of quantum mechanics could ultimately be derived by the proper interpretation and understanding of the results of the double-slit experiment alone. Furthermore, he discovered through a mathematical model that he devised that the interference patterns produced by the double-slit experiment, even the single-photon-at-a-time variant, could be accounted for by allowing each photon to simultaneously take every conceivable path through the universe and summing up the contributions for each path into a single, composite pattern.

In essence, this is precisely what my theory says. Let me explain.

The photon, which is simply a quantum of energy, is emitted from the lamp. Since we have not attempted any observation or measurement, the photon is in an indeterminate state. We perceive that this packet of energy is moving through space, and furthermore, we calculate that the packet is moving at the speed of light in the form of a probability wave. This form of energy is called potential energy, according to my theory. As such, the potential energy remains a component of the voidverse.

The voidverse manifests itself in the observerse as void, or empty space, from the perspective of our observational abilities. Empty space is a single, contiguous entity comprised of all the potential energy possessed by our universe, which is quite large but finite. In the form of an indeterminate probability wave, this quanta of potential energy is part of the single but greater composite pool of the universe's potential energy. This potential energy occupies the vastness of empty space that pervades every corner of the observerse; therefore, in its indeterminate form, this quanta of potential energy appears to an observer in the observerse to have passed through each point of the void's expanse on its way to our measurement apparatus. As we then perform a measurement, the quanta of energy collapses from its indeterminate, potential energy state to the determinate, "kinetic" energy state we observe.

If we constructed a model of the photon's movement in an attempt to comprehend the results we observe for the double-slit experiment, as Dr. Feynman did, we would out of necessity have to use the quantum mechanical description of a photon, which represents the photon as a probability wave function. This description of the photon is the only one we have that captures the potential energy essence of the photon during its stay in the voidverse. We have no other way of describing the photon in its indeterminate state, because it is not directly observable by any observer until the photon's energy is transferred from the voidverse to the observerse. This transfer occurs at the precise moment in time in our hemiverse at which we perform a measurement, forcing the quantum to shift from its potential energy state in the void hemiverse to the determinate state we observe in our observerse.

As you can see, my theory faithfully preserves the complete essence of Dr. Feynman's model.

Chapter 28

Big Bang, Not

My theory also suggests that science has really painted itself into the cosmic corner over its discussion of the origin of our universe.

Back in 1927 Georges Lemaître developed and published his "Hypothesis of the Primeval Atom." This hypothesis evolved into the theory known as the Big Bang, in large part, due to Edwin Hubble's observation, published in 1929, that the spectra of distant stars shifted in the direction that indicates movement away from us.[17]

Hubble performed terrestrial, telescope-based observations of numerous luminous objects within and far outside our galaxy and noticed a peculiar feature. The objects displayed light spectra that were different from what he had expected for the types of stars he observed. He noticed that the spectra were shifted in one direction, with the wavelengths of light shifted toward the red end of the spectrum.

This spectral shift to longer wavelengths immediately brings to mind the similar and well-known effect called the Doppler shift, named after the Austrian physicist, Christian Doppler, who first described the phenomena with respect to acoustic "sound" waves, which is a velocity-dependent effect that is the basis for the speed radars used by police. In similar fashion, Hubble compared the spectral shift of the distant stars and galaxies that he measured and compared the shifts against their distance from us. The distance to these objects could be determined fairly precisely by observing the change in apparent position of the star at different points along the Earth's path around our sun.

Simple geometry could then be used to calculate the distance to the object using the shift in its apparent position, or parallax. This geometric method for calculating distance results in units of parsecs which refer to *para*llax of one arc *sec*ond. Using this technique, one parsec is equivalent to a distance of about 31 $\times 10^{12}$ kilometers.

After comparing the distances calculated by parallax to the red shift observed in each star's spectra, Hubble discovered that a strong correlation existed between the two seemingly disparate measurements. His analysis of the data led him to interpret that everything beyond a certain distance from us was moving away from us. The further away a particular celestial object was from us, the faster the object appeared to be moving away from us. From these perceptions, Hubble concluded that everything in the universe is moving away from everything else.

But Hubble went even farther in interpreting his perceptions. In 1929 he believed that his observations, perceptions, and interpretations supported the notion that the universe must be expanding. This interpretation fit very neatly with the "Hypothesis of the Primeval Atom" that had been published by Lemaître only two years earlier, prompting Hubble and others to embrace the concept of the Big Bang as the most plausible explanation for the origin of our universe. In the end, Einstein acquiesced to the theory of the Big Bang, causing him to abandon the "cosmological constant," a fudge factor he had incorporated into the field equations of his Theory of Relativity to account for his belief in a static (non-expanding and non-shrinking) universe. In 1931 Einstein publicly abandoned his cosmological constant once and for all, referring to its inclusion in his theory as "the biggest blunder" of his life.

As you'll see in the chapter on dark energy, I believe that Einstein's abandonment of his cosmological factor was hasty and premature. In fact, I would go so far as to say that the biggest blunder of Einstein's life was the abandonment of the cosmological factor, followed only by his blunder in denying the "existence" of entanglement.

If I understand my own theory well enough, there are clear and compelling implications involving the origin of our universe. In turns out that any discussion of an "origin" to the universe should just be thrown out the window.

It goes like this. The voidverse possesses all of the universe's potential energy. Now, let's take the extreme case in which 100 percent of all the energy in the universe is in the form of potential energy—which, by the way, can never be the case and I'll explain the reason for this in the section covering the extreme ends of the energy balance spectrum. In the unrealistic case in which the universe's energy is entirely 100 percent in the void hemiverse, the only thing this situation could possibly represent is the state of a "potential" universe. In other words, if the universe's summed energy belongs entirely to the void hemiverse, there would be no observer hemiverse. There are at least three implications to this situation:

1. Without observers, there's no perception of a universe and therefore it does not exist, which is sort of the "nonexistentialist" perspective of "I don't think, therefore, I'm not."

2. The state of potential existence is not limited to universes, so anything that doesn't exist can be considered to potentially exist, and potential existence doesn't carry a whole bunch of clout, in my opinion. Therefore, although, the state of potential existence must necessarily precede actual existence, this alone is insufficient to become a card-carrying member of "actual" existence. The implications of this concept are that the usual criteria for determining the existence of something are not fulfilled. Even though the universe might occupy a state of 100 percent perfect potential, whatever the duration or extent of this state might be, the universe could not be perceived as existing. Therefore, a universe in which all of its energy resides and remains in the void hemiverse possesses nothing that can result in the fulfillment of the requirement for coming into existence.

3. Finally, and perhaps the least abstract of these terribly abstract implications, the concepts and dimensions of three usual dimensions of space and one usual dimension of time belong exclusively to the observer hemiverse. The consequence of this statement is that another concept of space and time appears to apply to the void hemiverse. If this is the case, then time as we know it is undefined for the duration or extent to which the universe occupies the state of 100 percent potential energy. Since the "existence" of the universe depends entirely, according to my theory, on the "existence" of both the void and observer hemiverses, the very concept of an origin to the universe remains undefined, since time does not apply to a universe comprised solely and exclusively of potential energy in the form of the void hemiverse. The clock only ticks in the usual way in the observer hemiverse, because that is the only place in our universe where it can. The clock ticks so long as the observer hemiverse possesses a non-zero energy value.

These arguments would appear, at first, to yield a cosmic version of the proverbial "chicken or egg" problem; however, that would be a false impression for reasons I expound upon in the section describing the spectrum of possible balances between the energy of the void hemiverse and the observer hemiverse. For now, let it suffice for me to say that the prospect of having a universe comprised of only potential energy appears to be forbidden, or at least highly discouraged. The balance of energy in the voidverse can only asymptotically approach 100 percent, but it appears to me that a universe in which the entirety of its energy is expressed as potential energy cannot "exist" in any sense of the word.

The first reason for this is, as I discussed above, that something in a purely potential energy state cannot be said to exist if there is no actuality or reality establishing its existence. But, equally important, I'll demonstrate to you that the void would appear to possess mass from the perspective of quantum statistics. The emergence of virtual and real particles in the void arising from quantum fluctuations limits the state of the universe to one in which the maximum potential energy is incrementally below 100 percent. As we discussed earlier, we can and do observe the emergence of virtual and real particles from the void.

As a result of this phenomenon, the energy possessed by the observer hemiverse is forbidden from having a zero value, and hence a ticking clock exists by which to attempt a perception and interpretation regarding the age or origin of the universe. For the reasons described in this section above, such an attempt may prove to be nothing more than a meaningless illusion. We can only calculate the age of the universe based on observations limited to the observer hemiverse, but the universe is comprised of two hemiverses and not just the observer hemiverse.

As you'll also see, either endpoint of the energy balance spectrum is deeply problematic for citizens of the observer hemiverse.

Chapter 29

The Extremes of the

Energy Balance Spectrum

L et's take what we've discussed and apply it now to the range of possible energy balance values between the two hemiverses. Sure, it's tempting to just say that the extremes of the energy balance spectrum are 100 percent potential energy (all of the universe's energy in the void hemiverse) and 100 percent "kinetic" energy (all of the universe's energy in the observer hemiverse), but it appears to me that these extremes are either not defined or not permitted. These degenerate extremes don't seem to represent universes, at least from the perspective of what we would typically consider to be a meaningful universe.

Let's start at one extreme and work our way to the other. We'll begin with the case in which the void hemiverse possesses the entirety of the universe's energy. In this case, the observer hemiverse possesses zero energy. We can comfortably assert that since the observer hemiverse possesses no energy, it possesses no matter; there are no observers, and there is nothing to observe. This is the equivalent of stating that the observer hemiverse doesn't "exist."

By the same token, if the universe has placed all of its energy eggs in the void hemiverse basket, then the universe is only "virtual" in the sense that it only *potentially* represents a universe. Without an observer hemiverse—no objects, no phenomena, and no observers—there's just no basis to establish the actuality or existence of the universe. So, in my mind, it's perfectly safe to say that a universe possessing only potential is really no universe at all.

It seems like a good idea to return to the concept of the vacuum, or quantum, fluctuation energy. The concept is often described in quite colorful terms as being akin to the surface of a violent sea, in which the zero-point energy (the one-dimensional, oscillation energy value of that particular spatial point in the void) fluctuates, or shifts, violently and unpredictably about some average minimum-energy value. In other words, the observed energy value of this point would sometimes dip below the value expected for a point in the void and sometimes the value would jump above the expected value. There's only so much the energy could drop below the expected value, because the observed magnitude of the value can't drop to or below zero. On the other hand, the observed energy value of that same point in space might be observed to shoot up for short spans of time to very high values, because there's nothing stopping that from occurring other than the diminishing probability of such an occurrence. Over some integrated span of time, an observer would measure the average energy value associated with that point of space as the value expected for that point in the void.

So, essentially, if the energy fluctuation is small, an observer would detect the appearance of virtual particles out of the "thin air" of the vacuum that come and go with the flow and ebb of energy at the vanishingly small dimensions of quantum phenomenon. On the other hand, it's possible for an observer to detect the "creation" of a real, honest-to-goodness particle. Of course, this should sound fairly mundane at this point in the discussion based on the theory I've proposed, since this observed fluctuation in quantum energy is really nothing more than energy flowing back and forth across the boundary separating the two hemiverses. It seems pretty fundamental and it must happen all the time.

The important thing here is that the observed energy level is heavily constrained in how low it can fluctuate, but the sky is the limit in terms of how high the level can fluctuate. Of course, small fluctuations are a dime a dozen and dramatically outnumber the frequency of high energy fluctuations.

Let's try and apply an analogy involving the process of evaporation, another quantum phenomenon that we discussed earlier. For this example, we're going to take an observational perspective outside the universe so we can view the contributions from each of the hemiverses. In this analogy, we've got what looks like a half-full glass of water sitting on the cosmic counter of our cosmic kitchen. In the bottom half of the glass, quanta of potential energy in the form of unobservable, indeterminate probability wave functions are swirling around like molecules of water. The upper portion of the glass does not possess the "water" of potential energy; instead it possesses the "air" of the observerse's kinetic energy. The surface between the "water" and the "air" represents the boundary between the two hemiverses.

It's possible in our analogy for molecules of "water" (morsels of potential energy) to burrow across the boundary where they join the "air" in the upper portion of the glass (transformation into morsels of matter). If they succeed in crossing the boundary, then the likelihood of returning to their home "water" is exceedingly unlikely. Instead, these rogue morsels of matter mix with the other morsels of matter, integrate into the society of matter morsels, and start families of their own in the form of atoms.

On the other hand, if the morsels of potential energy attempt to escape but fail to do so, an observer in the upper portion of the glass would see an "almost matter" morsel (virtual particle) begin to materialize and then disappear.

Perhaps you can see the implication arising from this fairly pathetic analogy. If not, it's okay, because I'll just tell you what I think the implication might be. The implication is that some combination of partial and full collapses of their associated probability waves accompanies the appearances of these virtual and real particles. The consequence is the collapse and transformation of some amount of potential energy from the void hemiverse into an equivalent amount of virtual and real matter in the observer hemiverse, along with the mass that necessarily accompanies the probability wave collapse, whether partial or full. An observer in the observer hemiverse would indirectly detect an ephemeral virtual mass to the void, along with a direct detection of matter, and an indirect detection of this matter's mass.

So, a universe comprised of pure potential energy (in other words, a universe in which all of its energy is stored in the void hemiverse) would spontaneously spawn its companion observer hemiverse containing, at least initially, a small concentration of matter. Nonetheless, as soon as any matter appears in the observer universe, the clock of time begins ticking and the spatial-temporal dimensions of the observer hemiverse would be established. Effectively, this scenario represents the effective endpoint of maximum potential energy. That is equivalent to saying that the minimum effective "kinetic" energy of the universe is just a tad above zero.

If we presume that what we have been taught about entropy is at least some- what true, once the observer hemiverse is established, entropy increases within this hemiverse, resulting in a somewhat steady flow of energy from the void hemiverse to the observer hemiverse.

Energy continues to flow into the observer hemiverse until the matter den- sity is sufficient for atoms of hydrogen to form and coalesce into stars. Once that happens, the evolution of the system naturally produces helium by fusing together hydrogen nuclei in these hot, stellar furnaces, and so on to heavier and heavier elements which populate the periodic table with which we are so familiar. Sometime after that point, life emerges and observers begin to inhabit the observer hemiverse. The subsequently increasing frequency of observations results in the continuing transfer of energy to particles as acts of observation force the collapse of indeterminate probability waves, followed by the subsequent transfer of energy back to the void hemiverse as particular observations cease.

The stuff in the middle doesn't seem terribly interesting to me, so let's just skip to the other end of the spectrum. In fact, let's go straight to the other extreme and work our way back to the middle.

The other extreme is a universe that stores all of its energy in the observer hemiverse. Such a universe, by definition, possesses no potential energy, and as we all know by now that means ... what? That's right, it means there's no empty space and it's time for a new entry in our glossary.

◊ **Black hole** – a volume of space in the observer hemiverse which possesses zero or near-zero potential energy.

This could be a problem for observers that depend on their bodies being comprised of atoms. In this extreme situation, there is no potential energy, and if an observer *could* provide an observation of the observer hemiverse, he or she might describe it as a bit cramped, since the observer would perceive no empty space. Of course, no empty space also means that there would be no atoms, since the particles comprising proper atoms need to be separated from one another, with a tight little nucleus in the center and a shiny outer shell of electrons.

Instead, no potential energy would force matter into a very hot, dense soup, or plasma, of determinate states that condenses down further to a solid, motionless block of "that which has no name" and not a single observer to notice. Such a universe might appear as the equivalent of the "mother of all black holes," but without observers it would also simply not be observable. As a result, I would be hard-pressed to call this a meaningful universe, even if it represents one possible fate for our own universe, although an unlikely one as I'll describe below.

Okay, so let's turn down the energy thermostat a bit and allow some potential energy to remain in the void hemiverse. In fact, let's leave sufficient potential energy for observers in the observer hemiverse to have a little elbow room. In this case, we have observers but they can't, as the saying goes, "swing a cat" without hitting an observable object or phenomenon. The universe has way more stuff in it and way less empty space than the universe we like to sidle up to today.

Now, let's turn up the thermostat again and examine the prospects. As we continue to transfer energy to the observer hemiverse, it begins to get hot and crowded until it gets to the point where observers consider it uninhabitable, and they begin to drop like flies. The ensuing reduction in the frequency of observations promotes the return of determinate states back into indeterminate probability waves, resulting in the net transfer of energy back to the void hemiverse. Of course, it would be difficult to definitely assert that such a thing *would* happen, but it appears from my theory that such a thing *could* happen. At the very least, such an outcome appears to me to be consistent with my theory.

In summary, the energy balance between the void hemiverse and the observer hemiverse appears to me to be constrained to stable values that exclude the extreme endpoints. It is only between these extremes that a universe can contain the necessary ingredients of two hemiverses and observers to influence the energy balance, just like the universe in which we find ourselves.

Chapter 30

Curved Space versus Flat Universe

If you're like me, the way in which most cosmology books banter about flat space, curved space, a flat universe, a positively curved universe, and a negatively curved universe is as confusing as it could be. I thought I'd try my hand at providing a simplified description of what the heck these things mean.

Let's begin with the shape of the universe. The possible options are (a) positively curved, (b) flat, or (c) negatively curved. There are no other choices.

Think of a positively shaped universe as being a cosmic bowl facing upward into which we rapidly pour all the stuff we observe in the universe. The stuff swirls all around; it flows up the sides of the bowl; it coats the sides of the bowl. Given the right conditions, the stuff settles in at the bottom of the bowl and, if gravity had its way and became the dominant force in the universe that it can only dream of becoming, all the stuff in the bowl would occupy the same point of space at the very bottom of the bowl. This upward-facing bowl represents a universe with sufficient mass to overcome the perceived force of the Big Bang that placed all the stuff in the bowl to begin with. Enough mass means enough gravity, and enough gravity means that, whatever happened during the Big Bang to get the universe's stuff all over the sides of the bowl, there's also enough gravity to bring it all back together again in the reverse of the Big Bang, equally ridiculously yet aptly named the Big Crunch.

The flat universe doesn't require much imagination. Think of the flat universe as a flat cosmic counter. Now do the Big Bang thing in your mind and pour the entire universe's stuff all over the counter. After the force of pouring the stuff onto the counter settles down, the stuff just sits there like a bump on a log. It doesn't coalesce back together like the fractured T1000 in *Terminator 2* (the scene involving the steel mill). The stuff doesn't spread further out. It just sits there and it's just gonna keep sitting there. Such is the analogy of the flat universe in which the mass of the universe is precisely right to create precisely the right amount of gravity to precisely cancel out the force of the Big Bang that supposedly started it all. The Big Bang Theory really snubs its nose at this possibility; you'll soon see that the forces promoting the concept of the Big Bang convinced Einstein to abandon his cosmological constant fudge factor in his Theory of Relativity that would have allowed such a flat universe condition to be established and accepted.

If you have the previous two concepts well under control, you might already have speculated successfully about the final option. Think of a negatively shaped universe as an upside down cosmic bowl. Pouring won't work too well in this situation, so think of injecting the stuff of the universe through a garden hose into the bottom of the upside-down bowl, which of course is now really the topmost feature of the bowl. The universe's stuff just flows out of the bowl and just keeps going. I guess I should have mentioned that our cosmic kitchen has no floor, so the stuff that fell out of the bowl just keeps going. In this analogy, the force of the Big Bang overwhelms the force of gravity produced by the mass within the universe. Gravity cannot stop the universe's stuff from continuing to expand outward. The Big Bang doesn't specify into what the matter actually expands, so the usual, utterly absurd answer from scientists is, "Well, it just creates new space, I suppose." Personally, I never found this answer particularly convincing or *ane* (the apparent antonym of inane or irrational).

So, if the theory of the Big Bang goes up in a Big Puff of Smoke, these concepts of the curvature of the universe go away, too, thank goodness.

Now, the curvature of space is a far more interesting topic of discussion because it's actually *ane*. Einstein's Theory of Relativity showed us once and for all that space is curved, but he never got around to explaining why it's curved.

How does one get his or her mind around the concept of curved space? Here's one suggestion for you. To understand curved space, let's use an analogy involving a familiar type of map, called a topographic map, or topo map (see Figure 10). A topo map shows the terrain of a region superimposed on the map's coordinate system, usually consisting of a Cartesian coordinate system using the region's longitudinal coordinates along the horizontal axis and the region's latitudinal coordinates along the vertical axis. The lines on the topo map show places of equal elevation. In terms of gravity, places with the same elevation possess the same gravity potential. So, in essence, a topo map displays lines of equigravitic potential superimposed on a Cartesian coordinate system. The map allows a viewer to "see" the curvature of the terrain with respect to the region's spatial coordinates. When we discuss the curvature of space, we are simply extending this concept into a third dimension. Please keep this analogy in mind when trying to understand the curvature of space resulting from the collapse of indeterminate probability waves down to particles possessing determinate characteristics. My theory proposes that mass is simply the quantification for the amount of curvature, or "puckering," of gravitic potential in three-dimensional space resulting from the local reduction in potential energy.

The first thing you probably noticed about the curving of space is that it only comes in two varieties—not three, like the shape of our universe. The choices are limited to (a) flat or (b) curved. That's it. Personally, I don't like too many choices, so I think this is the cat's meow. Like so many other things in life, it may not be as simple as it seems, but it's not too onerous either.

Let's start with curved space. Curved space is nothing more than space that possesses indirectly observable curvature as a result of the presence of mass and gravity. No mass, no gravity, no curved space. We observe matter and the manifestation of gravity in our universe. The range of gravity is not limited, although it drops off with the square of the distance. So the "existence" of any gravity anywhere in the universe curves space to one extent or another everywhere in the universe. If you hide a piece of space in a covered box and insist that I guess whether it's flat or curved, I know with absolute certainty that it's curved.

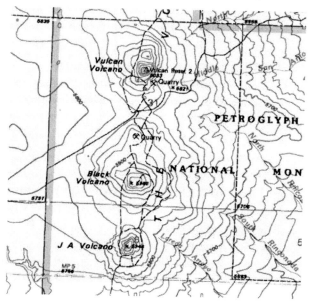

Figure 10. An example of a topographical map showing the equal elevation lines of the terrain (equigravitational potential lines) superimposed on the map's Cartesian grid of latitude and longitude lines. Map Source: U.S. Geological Survey.

Flat space, on the other hand, seems at first like an okay idea but doesn't actually "exist," for the reasons mentioned above. If you take a piece of space—any piece at all—in our universe, it's curved because of the related phenomena set we know as matter, mass, and gravity. Flat space is the equivalent of a perfectly flat Euclidean plane, and we already know that stuff like perfectly flat Euclidean planes don't actually exist in our universe.

So, then, you might be thinking, "How come space curves in only one direction, not two, like the shape of the universe?" That's a great question that has kept quite a few scientists going for some time.

Let me begin by suggesting that you can only find space curved in one direction and space curved even more in that one direction. In fact, I propose that there is a minimum curvature that space can possess which is set by the vacuum fluctuation energy. From our perspective, space can never, ever be completely empty. Something inevitably invades each part of space, such that it can be a darn restful place but never achieve a complete state of void serenity. No matter what you might try to do to empty a region of space of its contents, something will remain and/or something will appear to shatter the prospect of a perfect vacuum.

Quanta of electromagnetic energy might be traveling through. The ever-present and unavoidable gravitational field permeates all of space like a guest that will just never leave your side no matter where you go. This non-zero value of the gravitational field curves space everywhere. If you try to suck out all the tasty little morsels of matter, new ones will appear in the form of virtual or real particles. On the other hand, the most curved that space can become is represented by the concept of black holes, which possess nearly a complete set of determinate states and the near-total absence of potential energy. Everything else is somewhere in the middle.

So, then, why only one curvature to space? Everything we observe in our universe appears to contribute only to a one-sided curvature. Everything. No exceptions: matter, antimatter, energy, the four fundamental forces. Well, that tells me something. That tells me I perceive a truly fundamental property of the portion of the universe in which I'm an observer. It tells me the observer hemiverse can only manifest curvature of space in one direction. If I want to see space curve differently, then I'll have to find a way to become an observer in the void hemiverse, and that, put bluntly, won't be happening. When I combine this perception with my perceptions of Lorentz' transformations and Einstein's relativity, a funny thing happens. The dimensionality of the void hemiverse pops right out. Just like that. Please do read on.

Chapter 31

The Dimensions and

Dimensionality of the Hemiverses

Beginning with this chapter, I'm going to really pick up the pace and provide less commentary. The reason for this is simple. I've written 100,000 words in the three weeks since I started writing this book, most of it the old-fashioned way, using pen and paper. (This might appear quaint, but the reason is that I'm a terrible touch typist, although I improved my skills a bit during the writing of the second half of the book. In any event, I'm exhausted and I'm going to wrap the book up so I can enjoy Thanksgiving with my family and attend Roger Waters' performance of *The Wall* in Phoenix. I'll make every attempt to add commentary on my Web site and look into the prospect of writing another book. I already feel some things stirring around in the back of my mind!)

So, I'm going to break with my long-winded introductions and cut right to the chase on this one. The dimensionality of each hemiverse emerges as a set of philosophical implications arising from prior scientific measurements and the prudent filtering of scientific interpretation as I saw fit, based upon the image forming in my mind. I see only one way to assemble the pieces and remain fully consistent with scientific evidence. I'll drop additional dimensionality implications on you from time to time during the few remaining chapters of the book. In this way, I'll provide the rationale and context that validate the basis for my propositions.

The observer hemiverse (*observerse*) has four dimensions. Hopefully, if you chose to read this book in the first place, this does not come as a surprise. This is the part upon which most scientists agree, with the possible exception of the string theorists. So, just to be sure there's no ambiguity regarding the observer hemiverse, the observerse we occupy possesses three orthogonal dimensions of space and one dimension of time that together comprise Einstein's space-time continuum. As Einstein showed, there's an intimate relationship between the spatial and temporal dimensions that ensure a consistently observed value for the speed of light. I'll explain the basis of this intimate relationship later in this section.

So, an example of an observer hemiverse spatial dimension unit is the meter, represented by its abbreviation, m. An example of an observer hemiverse temporal dimension unit is the second, which is abbreviated as s. With these units, velocity is expressed as change in distance divided by change in time to provide units of velocity of meters per second, abbreviated simply as m/s. Acceleration is expressed as change in velocity divided by the change in time, which is abbreviated as m/s^2.

I propose that the void hemiverse (voidverse) also possesses four dimensions, but not just any four dimensions. The four dimensions of the void hemiverse are comprised of three inverse spatial dimensions and one inverse temporal dimension. So, the void hemiverse possesses three orthogonal dimensions of inverse space and one inverse dimension of time that together comprise the space that is entirely unobservable to us. These unobservable, tightly compacted dimensions should make the string theorists squeal with the delight of their vindication. The various equivalent string theories comprising Edward Witten's M-Theory may represent nothing more than the different perspectives from which one may view the universal scene. I'd like to make one last super-speculative comment about the implication from String Theory that the universe appears to possess eleven dimensions. Perhaps these represent the eight combined dimensions of the two hemiverses and the three dimensions of the surface that contains them, as in the "actual" outer envelope of the universe.

So, an example of a voidverse spatial dimension unit is the inverse meter, represented by its abbreviation $1/m$ or m^{-1}. An example of a void hemiverse temporal dimension unit is the inverse second, which is abbreviated as $1/s$ or s^{-1}. With these units, velocity is expressed in the void hemiverse as change in inverse distance divided by change in inverse time to provide units of velocity of seconds per meter, abbreviated simply as s/m. Acceleration in the void universe is expressed as change in inverse velocity divided by the change in inverse time, which is abbreviated as s^2/m.

Please recall that earlier in the book I mentioned that there were scientists back in the 1800s and the early 1900s who asserted that electromagnetic waves required a medium for its propagation through space, and this resulted in the creation of the concept of the Luminiferous Aether? Perhaps you also remember that I mentioned that the science community sort of lost interest in the concept after Einstein's Theory of Relativity arrived on the scene. You might also recall that I mentioned that the Theory of Relativity mathematically described the way in which light propagated through space, but the model didn't provide insight into the physical nature of the medium that carried these waves. Well, the Luminiferous Aether wasn't some delusional concept after all. It exists; it's called the voidverse, or the "empty" space that we're unable to directly observe but actually possesses physical properties. Scientists who supported the notion of the Luminiferous Aether had compiled a long list of physical attributes it must possess. If you feel so inclined, you might want to investigate the matter further on your own and see how well those attributes apply to the inverse space-time of the voidverse.

So, an implication that pops out of this interpretation is that space-time is smooth and continuous. Space-time cannot be broken down to individual quanta of distance or time. There's nothing equivalent to a fundamental "particle" or minimum unit of time, and time is not comprised of degenerate extents of time embodied by the term "moment." In the same way, there's no fundamental "particle" of space, and space is not comprised of degenerate extents of space embodied by the term "point."

It would seem to me that the notions of points of space or moments in time arise solely from our perception of particles representing fundamental bits of matter. This analogy, then, seems to represent an attempt to convey the concept that developed from our perceptions of vanishingly small bits of matter we call fundamental particles to our notions of space-time, and there appears to be no sound basis for doing so. Thus, the concepts of moments and points, in essence, represent another attempt by our minds to establish the "existence" of infinity, which doesn't appear to exist whatsoever in the universe. Points and moments represent nothing more than another illusion upon which to construct paradoxes.

Chapter 32

Gravitational Waves

First of all, there are gravity waves and there are gravitational waves, and most cosmologists would throw a condescending little snicker your way if one carelessly applied the improper term. So, for starters, we'll provide a short lesson in cosmology etiquette before tearing apart the concept embodied by each term.

Gravity wave is the fancy scientific term for the type of wave that we're accustomed to seeing. When your child drops small rocks into a pond, delights in the plunking sound they make, and then says, "Mommy/Daddy, look at the pretty waves," the little tyke actually means "Mommy/Daddy, look at the pretty [gravity] waves." I'm not necessarily suggesting that you correct them unless, of course, you really want to indoctrinate them early into the current culture of science and risk them having a less meaningful childhood experience.

On the other hand, gravitational waves are a very serious cosmological matter and should not be confused with childhood notions of waves. Don't say that I didn't warn you.

The irony, of course, is that there are more similarities between them than there are differences, except with respect to the "hard" science thing.

Okay, now we're ready to split the concepts down to their fundamental meanings and build up something we hopefully consider useful.

Let's start with the less controversial gravity waves. The term "gravity wave" emphasizes the role that gravity plays in the formation and propagation of waves along the surface of water. The surface of the water actually represents the boundary between the water beneath it and the ocean of air above it in which we find ourselves immersed and live out our daily lives. The large volume of the air above the surface of the water represents compositely a large amount of mass and the weight of this mass exerts a meaningful force against the surface of the water that we call air pressure. At sea level, the force of this pressure is about 14.7 pounds per square inch, a value for which several different standards exist. Above sea level there's less air and hence a lower air pressure; below the elevation corresponding to sea level the air pressure is higher. Of course, if you're actually below water, you'll observe a much higher pressure—technically, that's not air but water pressure.

The air is applying pressure to the surface of the water and the surface of the water is applying pressure to the air. As we discussed earlier, the process of evaporation allows energetic molecules of water to cross the boundary and enter the air as molecules of gas (steam) that then become part of the ocean of atmosphere enveloping our planet. The competition between these two forces (water pressure exerted upward and air pressure exerted downward) is the cause of the gravity wave phenomenon we observe.

When we drop a stone into the water, we force the air–water interface (the water surface) downward under the added force of the falling stone. From this point onward, the interface between the air and water behaves as a pendulum. First, the water pressure fights back at having been displaced downward by forcing water back into the volume that was displaced. The momentum of the upward surging water then crosses the original, stable level at which it started by surging further upward into a "metastable" condition that cannot be sustained against the force of air pressure. Not to be undone by the water that lives beneath it, the air fights back at having been displaced upward by forcing air back into the displaced volume, which results in a metastable condition in which, once again, the water surface has been displaced below its stable level.

Neither of the metastable conditions can be sustained indefinitely, so we watch the water surface bob up and down as though it's a pendulum which loses some of its momentum with each swing; that is precisely what it is. As the energy introduced by the falling stone starts the pendulum swinging, the resulting waves spread outward in an ever-growing circle and the energy dissipates into a larger and larger area. Since the circumference of a circle is proportional to its radius, the strength of the waves diminishes with distance, not the square of the distance, as is the case with a force that expands outward in three spatial dimensions. The strength of the waves therefore drops off with distance. Eventually (in time and space), the momentum transferred to the water surface by the falling stone dissipates and the surface of the water re-attains its Zen-like, smooth quality, so we can more easily see the look of enlightenment on our faces being reflected back at us.

If the water and air were constricted as in a piston-like configuration then the pendulum motion would be limited to only up and down. However, the force of the original displacement also spreads out to the sides and propagates outward until it reaches another surface or boundary that inhibits its motion. Each point along the undulating surface of the water can be thought of as a pendulum swinging under the dynamically changing conditions resulting from the competition between the forces due to air and water pressure. Without gravity, there would be no gravity waves.

Needless to say, without gravity there also would be no gravitational waves. So, it's time to wipe the Zen-like smiles off of our faces and get down to the more serious business of gravitational waves.

For the moment, let's forget about whether mass curves space (as that guy Einstein suggested) or the curvature of space causes mass (as I propose). A gravitational wave represents the concept that dynamically changing gravitational conditions in some region of space propagate outward through space altering the curvature, or spatial dimensions, of space. Over time, the metastable local gravitational fluctuations dissipate and the "normal" usual gravitational balance is restored if the cause of the fluctuation ceases, as in a one-time event akin to the dropping of the stone into the pond. Of course, the balance is not re-established if the gravitational fluctuations are caused by the cyclical rotation of one really massive celestial body around the other. This type of situation represents a gravitational wave generator. In essence, either situation can once again be represented by a model using a pendulum.

Let's start with the traditional model presented by "the other guy." Furthermore, let's choose a set of bounded circumstances to consider. Hmmm, we can choose either the former case represented by dropping a stone into a pond or we can use the latter case, which is somewhat similar to a wave pool at a water park. It really doesn't matter which situation we select as long as we're consistent. So, let's choose the former case and let's establish as our assertion that there's a huge change in local gravity arising from some cataclysmic event, like a supernova. The colossal explosion forcefully reduces the amount of mass and gravity in some local region of space but not before the mass dissipates, forcing our pendulum to begin its motion.

The range of gravity is not limited in space but, as we already learned, the magnitude of gravity diminishes with the square of the distance. In the situation I proposed, energy is released as a result of the transition from the bloated "before" local mass and gravity condition to the relatively emaciated "after" local mass and gravity condition, sort of like a cosmic Jenny Craig advertisement.

Now let's apply the pendulum analogy again. Relativity proposes, in essence, that the pendulum is comprised of the local gravity contribution pushing against the cosmic gravity background contribution on one hand and the cosmic gravity background contribution pushing back against the local gravity contribution on the other hand. (In all honesty, I prefaced my assertion with the words "in essence," because I'm trying to reduce an insanely complex set of conditions into a simple, succinct analogy—that is to say, one which heavily favors my proposition. I'm fairly confident that if any gravitational physicists read my book, I'm going to get lengthy letters of clarification and rebuttal, but I'm sticking to this analogy nonetheless.) The value of the gravitational field at some point in space will bob up and down due to the buoyancy between these competing gravitational forces. The rest of the mass in the universe will attempt to exert its gravitational influence on this point in space while the svelte, new, mass value of local space tries to exert its influence to produce a new balance point.

Since gravity exerts its influence in all directions, the waves propagate outward in space alternately squeezing and expanding space first in one direction and then in the other, akin to combining the effects of two undulating water surfaces: one placed vertically and the other placed horizontally at right angles to the first one. Since the surface area of a sphere is proportional to the square of its radius, the strength of gravity diminishes by the square of the distance.

The resulting gravity waves know no bounds but the usual, embarrassingly puny value of their strength drops rapidly with distance as its wave front extends outward in a growing sphere. As the wave front of the gravitational waves passes us, scientists hope to detect its influence by measuring tiny fluctuations in spatial dimensionality, using—you guessed it—electromagnetic-based observation techniques.

Now, let's migrate to the model I'm proposing in this book. Let's return in our discussion to the point where we asserted that there's a huge change in local gravity arising from some cataclysmic event, like a supernova. The colossal explosion forcefully reduces the amount of mass and gravity in some local region of space but not before the mass begins to dissipate, forcing our pendulum to begin its motion.

This picture might at first seem identical, but the events unfold in a very different way. Sure, a bunch of matter is ejected outward from the explosion and dissipated into a growing sphere until the matter density reaches the universe's average mass density level, or the matter gets trapped by some other celestial body's local gravitational field. But there's a whole other bunch of matter that gets converted back into energy and, as I've already described, energy can flow both ways across the barrier separating one hemiverse from the other.

My proposition is, as opposed to the assertion of the other guy, that the curvature of space is in fact formed by the coalescence of potential energy into matter that puckers space, creating the curvature that the other guy so completely and eloquently described. Furthermore, I contend that we perceive the resulting curvature as the property we call "mass." If the "kinetic," actualized energy comprising an individual particle is instead transferred back to the void hemiverse from whence it came, the energy it represented is moved back into the potential energy column. As this happens, space unpuckers, the corresponding curvature is relaxed, and the perceived mass disappears to any observer unfortunate enough to be that close to an exploding supernova.

Please allow me to clarify another way. The all-important surface that undulates to represent the bobbing pendulum is not observable in our hemiverse. I propose that the bobbing surface is instead the barrier separating the two hemiverses—as the forceful transfer of energy bounces this way and that, the barrier seeks its new state of balance and the swinging pendulum is eventually stilled.

What might an observer placed a safe distance away perceive under this set of circumstances? I'd be delighted to let the theoretical computational physicists and mathematicians have their say about this once the proper mathematical models have been constructed, but I'm willing to speculate just a tad.

First, the change in gravitational force would likely appear far less dramatic to a distant observer. Let's return for the moment to the analogy of the tree falling in the forest. Sure, the farther away the observer is placed from the falling tree the quieter the perceived sound will be. But let's modify the situation a bit to make it more analogous in my mind to the situation I'm proposing involving the exploding supernova. We're going to modify the falling tree scenario by disallowing the kinetic energy that forms as a resulting of the tree falling from being transferred to the ground, but we're still allowing the change in potential energy resulting from the tree shifting from its upright position to its position lying on the forest floor.

In this variant of the falling tree scenario, the energy of the falling tree is transferred to somewhere other than to the ground. We're allowing the kinetic energy instead to transfer between the observer hemiverse and the void hemiverse. It's not a perfect analogy I recognize, mostly because there's no analogy in our everyday observational world. The fact is that there's nothing from our observational lives precisely like the directly unobservable phenomenon I'm proposing to occur in our hypothetical situation.

In any event, the transfer of energy between the observer hemiverse and the void hemiverse is akin in its effect to gently lowering the tree into its new prone configuration. Sure, the potential energy of the tree has been changed but without the accompanying cataclysmic BANG.

Since our efforts to perceive gravitational waves are focused on detecting the BANG, there may not be anything extraordinary to detect. Let's break down the process whereby the gravitational wave reaches our instruments according to my proposal. First, some portion of the event's energy transfers back to the void hemiverse, so the magnitude of the observable effect is diminished. Next, the wave travels great distances to reach our neck of the woods, and the strength of the wave attenuates with the square of an enormous distance. Then, we're going to implement an indirect scheme as the basis of our measurement, which relies on the vanishingly small variations in space occurring in the path of photons we generate for this purpose. Finally, the strength of the wave needs to be sufficient for our apparatus to detect and measure above the background noise our universe routinely and obnoxiously generates. And, if that's not enough of a challenge, add into the mix the prospect discussed earlier of local mass density fluctuations arising from the presence of unobserved, albeit potentially observable, matter.

You may be curious enough as to inquire to the current state of gravitational wave science, so let me fill you in. In 1993 the Nobel Committee awarded its coveted prize in physics to Russell Hulse and Joe Taylor for their mathematical studies in gravitational waves based on observations of the Hulse-Taylor binary star system. Their models accounted for the apparent radiation of, and subsequent reduction in, gravity accompanying the collapse of the orbits of the Hulse-Taylor binary star system, one of which is a pulsar. The mathematical work was impressive, no doubt about it. The only problem is that no gravitational waves have ever actually been observed. Granted, the attempts at detecting gravitational waves are still young and somewhat immature, but certain aspects of the results are nonetheless quite interesting.

The most ambitious experiment to date is LIGO.

LIGO has been performing observations actively now since 2002, and there's nothing interesting to report yet. Well, not exactly. LIGO reported its first discovery on August 12, 2010. Unfortunately, this discovery had nothing to do with the detection of gravitational waves but, rather, involved the discovery of a radio pulsar using LIGO data analysis tools on data from radio telescopes rather than data from LIGO interferometers.

Keep in mind that hunting gravitational waves requires the right timing, just like the timing needed to perceive the sound of a falling tree. If no tree falls while you're listening, there's not much to observe. Gravitational waves that are strong enough for the equipment to detect are not everyday occurrences, so the absence of detection is not sufficient in and of itself to conclude that gravitational waves can't be observed.

It's only fair to say, in fact, that the results are thus far inconclusive. The strength of a gravitational wave signal by the time it reaches the apparatus has been predicted to be about one part in 10^{20} parts. That's 1 part of spatial fluctuation signal out of 100,000,000,000,000,000,000 parts of space; therefore, the measurement requires an outrageous level of sensitivity and precision. The funny thing is that, based on the likelihood that a proper cataclysmic gravitational event should already have occurred, the results from LIGO-based observations effectively place the upper bound of the gravitational wave signal nearly one million times smaller than current theory predicts. Oops.

Then there's the argument that holds for the electric fields associated with an atom's outer shell of electrons. Close up, the electromagnetic field value is palpable. Place some amount of distance between you and the atom and the electron shell's impact becomes negligible. Like Vinny in the movie, *My Cousin Vinny*, it just "blends" so you can just forget about its contribution. The analogy only works when you consider that close up, as in the small space of a courtroom, Vinny obviously does not blend. In fact, he sticks out like a sore thumb. Yet, if he were standing among a throng of people viewed from a distance, his outlandish-clothing field would become virtually or actually unobservable.

So, the strength of gravity, like electromagnetism and outlandish-clothing-ism, drops substantially with distance. It seems reasonable to me then that the prospect of gravitational waves being detectable and measurable over huge distances is about as unlikely as measuring the strength of an atom's electrons at an appropriately scaled distance. It doesn't really matter whether the gravitational wave forms as the result of a cataclysmic event or as a consequence of two disproportionately massive bodies orbiting each other: at some distance the signal will just reach background levels. That distance is probably less than the distance separating us from the celestial event we're desperately but fruitlessly trying to observe.

There's another aspect to recent observations involving gravitational fluctuations. Astronomical planet hunters use an effective scheme to search for and locate planets orbiting distant stars. Rather than trying to observe gravitational waves as a scheme to identify disproportionate masses orbiting each other (as in the case of a planet orbiting its star), these astronomers exploit the tiny shifts in the electromagnetic spectrum of the star. Two disproportionate masses orbit each other around their combined center of gravity. So, although we normally think of planets orbiting their lazy-ass star, both the star and the planets actually orbit each other. Since the mass of the star is colossal compared to most planets, it would appear that the planets are doing all of the moving and the star just sits there with its sunny grin.

In fact, the star makes subtle but perceptible shifts in its position as the planet, or planets, revolves around it. These positional shifts manifest themselves as Doppler shifts in the electromagnetic spectrum of the star. As the star moves away from us, the spectrum shifts toward the red end. As the star moves toward us, the spectrum shifts to the shorter wavelengths of the blue end of the spectrum. These spectroscopic measurements are combined with observations of the fluctuations in the star's brightness to estimate the diameter of the planetary object blocking the star during the non-shifting portion of the observation. (At the moment when both the planet and its star are aligned along our line of sight both objects are moving orthogonal to our line of sight, so no Doppler shift would be observed.) The spectral shift provides information about the planet's mass, and the brightness fluctuation provides information about the planet's size. From these observations, scientists then calculate both the density of the planet and its distance from the star to make a judgment call as to whether the planet should be considered "habitable." With improvements in measurement technology, it becomes possible to detect less massive planets. Scientists are already beginning to find the first "Earth-like" planets.

The point here is that we can readily observe subtle gravitational fluctuations using electromagnetic-based observation schemes. Gravitational detectors don't so much as blink in response to the low gravitational fluctuation levels with which planet hunters routinely deal.

You might expect gravitational physicists to be deterred in their endeavors and take pause to rethink the whole current theory through a bit more. But, as is pretty common in the current industrial-like culture of science, an inconclusive experiment can mean only one thing: It's time for an even bigger and far more costly experiment. "Damn the torpedoes, full speed ahead!"

LIGO is about as big as we can go on the surface of a somewhat smallish, curved planet, since we need two precisely straight optical pathways at right angles to each other in which to shine our photon-based measurement systems. There are other terrestrial detectors that should improve the sensitivity of such schemes by a factor of ten and extend the volume of space that the instrument can effectively probe by a factor of one thousand, or 10^3, since there are three spatial dimensions. The next planned crop of detectors, such as the Laser Interferometer Space Antenna (LISA) and the DECi-hertz Interferometer Gravitational Wave Observatory (DECIGO) will thus be deployed far above the ground and the price tags will likely rise to match.

Chapter 33

The Fate of the Universe

Hangs in the Balance

Actually, the implications of my theory with respect to the fate of the universe seem pretty simple and straightforward. It would seem to me that analogies involving swinging pendulums are more appropriate than I had initially imagined. So, the fate of the universe actually might hang in the balance of the pendulum representing the boundary between the two hemiverses.

This discussion is going to be shorter than I could have possibly imagined even at the start of this book when I recognized with near-catatonic fear that this particular discussion could only be avoided for just so long. And, here we are.

In an earlier section dealing with implications of my theory, we discussed that certain conditions near the ends of the energy balance spectrum are prohibited. Specifically, the situations representing a universe comprised of pure potential energy and a universe comprised of pure "kinetic" energy are not permitted because of certain aspects relating to the quantum world. Okay, I still feel this way but with a twist.

Just as the snapshots of Zeno's moving arrow fail to capture the essence of mass and momentum in the dynamic situation of an arrow zooming toward its target, the consideration of the extreme ends of the spectrum represent pictures from which a key detail is missing.

The endpoints of the energy balance spectrum are not stable. First and foremost, that means that these extreme conditions cannot be sustained in a static, unchanging equilibrium. As we discussed earlier, such a set of conditions represents a metastable condition. This means that the conditions can be sustained for only an "instant" of time before the circumstances change in a dynamic way. Any attempt to contemplate the energy balance endpoints as snapshots fails to include the all-important yet fundamental nature of momentum. From the perspective of momentum that we've gained through the concepts I proposed in this book, we can contemplate the endpoints from a different perspective, one that now includes the role of momentum.

That's not the same as saying that the extreme endpoint conditions of the energy balance can't be achieved; it only means that the extreme conditions at either end can't be sustained. From our discussion of gravitational waves, I introduced the prospect that the barrier separating the two hemiverses itself possesses the attribute of momentum. As a consequence of the momentum this barrier can possess, a pendulum is formed between the dynamic gravity conditions presented by the exploding supernova. I further proposed that since at least some energy of the explosion is returned to the void hemiverse in the form of increased potential energy and the accompanying appearance of additional empty space (or relaxation of space curvature), the barrier between the hemiverses undulates as a pendulum until the event's energy smoothes out over the growing extent of space and time, at which point the hemiverse barrier establishes its new state of balance.

The analogy of a pendulum, of course, also means that momentum is a property of the barrier separating the hemiverses. If we apply this concept to our energy spectrum, we get some surprising results.

Even though the endpoints of the energy balance spectrum are metastable, the hemiverse barrier might possess sufficient momentum to carry it into the forbidden territory very near an extreme endpoint. In fact, the momentum might even be sufficient to take it all the way to the extreme end. And one very clear implication of my theory is that it just can't stay at that particular place. No way, no how.

So let's take another step back and set up our thought experiment. The initial conditions for this experiment are that the energy balance between the two hemiverses is swinging toward the end of the spectrum represented by a universe comprised only of pure "kinetic" energy, which an observer in the observerse would perceive as the condition represented by the complete and total absence of potential energy and empty space. This is a special observer, obviously, since the observer can hypothetically withstand all conditions but the final, extreme condition. That extreme, unstable, final condition represents a universe comprised of one hemiverse (the voidverse) with a zero energy state and the other hemiverse (the observerse) possessing the full, finite energy of the universe.

Now, we let the scene unfold. The momentum possessed by the pendulum (the barrier between the hemiverses) is sufficient to drive it all the way to the end of its swing, which in this case represents a universe comprised of one non-zero hemiverse. (The other hemiverse possesses absolutely none of the universe's energy at this one particular snapshot in time.) In fact, the momentum of the barrier achieves its own zero value at this point, but the condition is metastable and unsustainable. At that one and only one set of circumstances, the universe is comprised of one hemiverse and the observer has lost its observer status since he or she was obliterated as atoms devolved into an unimaginable dense solid state comprised of no empty space. "Immediately" following this snapshot frame of the universe, quantum effects require that some of the energy of the universe must "tunnel" to the other hemiverse, so a new barrier is established and quanta of energy flow back across it to what had once been the "void" hemiverse.

Here's where it gets really strange. As the pendulum now begins its return voyage, the only hemiverse in which an observer can possibly exist or emerge is the one we had been formerly calling the "void" hemiverse. The situation inverts, the names of the two hemiverses are exchanged and the whole darn process starts over again. In other words, an observer placed in the "hemiverse formerly known as the void hemiverse" would observe a universe devoid of matter and chock full of empty space. With the return swing of the pendulum, a complete, thorough condition of 100 percent "kinetic" energy becomes the equivalent of the complete, thorough condition of 100 percent potential energy to an observer in the other hemiverse. Isn't that wild?

Momentum is first imparted to the pendulum on its reverse swing by the inevitable and energetically favored appearance of virtual and real particles in the new observer universe. As the matter density of that hemiverse increases, atoms of hydrogen form, stars coalesce, atoms of heavier elements are produced by the stars, observers emerge, and observations cause indeterminate quanta of energy to take determinate values, further hastening the flow of energy from the new void hemiverse into the new observer hemiverse, and the pendulum picks up momentum once again.

It seems to me that the momentum imparted to the pendulum can achieve only one of two possibilities: either the momentum becomes sufficient to carry it all the way entirely to the other endpoint or it doesn't. If the momentum carries the pendulum to the extreme endpoint of the energy balance spectrum, then the hemiverse we now occupy collapses under its own mass and spawns a new observer hemiverse in what we had considered the voidverse, adding a great deal of forcefulness to the concept that the grass is greener on the other side of the street.

The consequence of this portion of the thought experiment is that the void hemiverse we now know leaves "existence," as we normally consider it to be, for a vanishingly brief span of time and space before it re-exists in the form of a new observer hemiverse.

In fact, the prospect of this implication alone sufficiently, fully, and uniquely defines the dimensions of the void hemiverse.

Now, let's briefly consider the situation in which the pendulum doesn't possess sufficient momentum to carry it all the way through its swing to the extreme energy balance endpoint.

Quite frankly, I don't feel good about this situation. It just doesn't sit right with me. First, I'll describe the way in which I think this situation would manifest itself, and then I'll explain more about why I don't like it.

In this situation, the barrier loses its momentum before it reaches one end point or the other. The barrier would in effect become something akin to an aimless bum, with energy flowing arbitrarily between hemiverses. The universe would appear static and arbitrary without any prospect of closure. The universe would become an infinitely dull place without the appearance of any meaningful changes in spatial dimensionality or the flow of time. This notion bears some resemblance to the "C-field" Theory of British astronomer and mathematician Sir Fred Hoyle.

Something Einstein said is ringing through my mind: "God is subtle but not malicious." In essence, I believe Einstein meant that all of what we perceive as nature must be based upon a simple set of consistent rules, because God is neither malicious nor arbitrary. In fact, I believe that it is precisely upon this simple basis that paradoxes emerge. Simply put, if you don't have accurate and complete insight into the rules of the game, lots of things will seem arbitrary, inconsistent, and paradoxical. This concept should sound very familiar by now.

Anyway, a universe that settles into an arbitrary energy balance state seems to me more than just a little capricious and arbitrary. I don't feel that I could offer a reasonable or meaningful explanation as to how or why the pendulum could or would lose its momentum, so although this situation has to be considered for the sake of completeness, it doesn't seem likely or right to me. I wouldn't expect the cosmic pendulum to just stop. Instead, I view the energy balance as though it were a yin and yang emblem spinning relentlessly and perpetually about its center: steady, sure, and constant. Like Einstein, I'm loath to the thought of a spinning yin and yang symbol stopping arbitrarily in its rotation like the wheel on *Wheel of Fortune* with the result that the contestants (we) get stuck with whatever "prizes" represent the arbitrary stopping point of the wheel. There is simply no way that I intend to live in any universe for which the Wheel of Fortune represents an accurate depiction. Forget it.

It's possible, I suppose, that the pendulum could come to rest at an equilibrium point at which the energy of the universe is split equally between two static hemiverses.

Having said this, I believe the universe is inclined to a state in which the pendulum is in constant motion. Although there may be circumstances that might be capable of altering the barrier's momentum, I think it's far more likely that the momentum of the barrier possesses a fixed, rigid, conserved value. I further believe that this value underlies and defines the cosmological values for our universe.

Sure, I've repeatedly asked you to imagine the barrier separating the hemi-verses as a pendulum, but I think there may be an analogy that provides a better image for some people. Remember, the position of our cosmic pendulum at any time represents the state of the energy balance between the void's potential energy and the observer universe's "kinetic" energy. Typically, when you think of a swinging pendulum the only thing you really observe is the pendulum's kinetic energy. We understand that at the moment the pendulum reaches the extent of its swing, it comes to rest for a moment during which its kinetic energy value is zero and all of its energy is now in the form of potential energy. So, if the pendulum image works for you then, by all means, stick with it.

This might actually introduce unnecessary complications, so just skip this paragraph if it begins to mess you up. Another way to think about this balance is to do just as I suggested a few paragraphs back. Imagine a simplified yin and yang symbol that spins at a steady rate of rotation. Think of a disk that can rotate about its center. Perhaps you stick a mental pin through the center of the disk and pin in up to an imaginary wall. Draw a straight line representing a diameter across the disk so the line crosses through the very center where you inserted the pin. Now, color half of the disk blue and the other half red. (Feel free to use different colors, if you want to make it more aesthetically pleasing to you.) Dangle a string so it hangs downward in front of the disk and visually slices through the center of the disk. Now, start the disk rotating at a slow, steady rate of rotation. Either half of the disk shows a continually varying mixture of red and blue (or the particular colors you chose) that represents the appearance or manifestation of the barrier's steady momentum. However, in this case, the dangling string represents the barrier separating the hemiverses, and the steady angular momentum of the disk represents the fixed momentum of the barrier itself. Again, just ignore this description if it misses the mark for you.

So, the bottom line I propose for the fate of the universe is this: The aspects of our universe are defined by the position and momentum of the barrier separating the void hemiverse from the observer hemiverse. The forces that contribute to the momentum of the barrier are sufficient to ensure that the pendulum reaches the extreme endpoints of the spectrum of possible energy balance states. When the pendulum reaches the maximum 100 percent "kinetic" state, the energy of the universe is conserved but the universe spontaneously shifts into the only alternate state available to it.

At the moment the phenomenon known as "particles" appear in what had previously been the void hemiverse, the designation of the hemiverses inverts. The consequence of this "hemiverse reversal" is that it's now the turn of "the hemiverse formally known as the void hemiverse to try its hand at being the new observer hemiverse. And the potential energy stored in what is now the new void hemiverse (but had previously been the observer hemiverse) starts manifesting itself in the new observer hemiverse as observable objects and phenomena, as well as the directly unobservable phenomenon manifesting itself as empty space.

Eventually, the table turns again and again and again. I'm sorry to disappoint you, but I'm at a complete loss as to how the table-turning itself ever started or how the table-turning itself might ever end.

Nonetheless, the impact of this proposition on our perception of our universe and our understanding of time is both staggering and starts right around the corner on the next page.

Chapter 34

The Flow of Time

I really hope that Stephen Hawking is reading this chapter. (I could really use a foreword for a future revision of this book. Hint, hint.) I've inducted Stephen Hawking into my personal gallery of role models from which I try so hard to choose the actions I take in life. I don't succeed as often as I would like, so I tend to come across a bit less like Stephen Hawking and more like Gregory House, M.D., but at least I try. If I were forced to provide a short list of my role models, it would consist of Albert Einstein, Stephen Hawking, Charles Ferri (my high school AP Chemistry instructor), Jean Luc Picard (of course), and my wife, Janna. In the event my wife is actually reading this book after all she endured during its writing, I would like clarify that I did not necessarily present the list in the right order. Okay, honey?

Stephen Hawking obviously spends as much time thinking about the nature of time as I do. He wrote *A Brief History of Time*[18], a wonderful book on this topic that I read and thoroughly enjoyed. The questions he posed in his book led, in part, to the thinking that in turn led to the writing of this book and the concepts contained herein.

So, without further ado, please allow me to propose a basis for the flow of time. Once you have the concepts I presented earlier assimilated into your thinking, the answer sort of just pops out (or so it seemed to me).

Let's set up our thought experiment, once again from the perspective of an observer in the observerse. The energy barrier pendulum is about to complete its swing to the extreme state of the universe in which the last of the potential energy will be transferred from the voidverse to the observerse. The last of the empty space is collapsing, the speed of light is slowing down due to the rising steepness of the minimum curvature of space, and the observerse is becoming very crowded as the last bits of empty space collapse into matter. Nonetheless, the momentum of the energy barrier pendulum is carrying it to the end of its swing.

There! The pendulum strikes one of the ultimate ends of its swing. Observers are nowhere to be seen. Of course, there's nothing to be seen. The observerse has been decimated into a solid block of matter that will soon fuel the conversion of the voidverse into the new observerse. There is no longer any such thing as an observer in what had been our observerse. In fact, there won't be another observer until the universal gravitational constant enters the "hospitable" window in the new observerse.

Now, we need to shift our view of the situation to that of the "God" perspective standing outside of the universe. What we'd observe is the pendulum momentarily frozen in its path as quantum fluctuations in the voidverse begin to manifest themselves as morsels of matter, and the pendulum begins the next in its relentless series of swings. The hemiverse that used to be our observerse begins to transfer its energy to the other hemiverse as though it were nutritious fertilizer being placed on the barren fields of the new observerse to nurture the new crop of observers that will inevitably sprout when the spring rains arrive in the form of a "hospitable" universal gravitational constant.

As this begins to happen, the dimensionality of our space-time inverts. The prospect of any observers arising under such conditions in what has become the new voidverse seems unlikely at the very best. Sure, the second law of thermodynamics could be invoked and it would not be inappropriate to consider that the entropy of our hemiverse is flowing in reverse, but I don't think that's an appropriate analogy, since space-time inverts and doesn't change its sign. After all, the magnitude of space-time is changing and not its direction or polarity. I think that the second law of thermodynamics may not even be correct or applicable here. It may be the result of a faulty interpretation arising from accurate measurements. I think it may be the result of nothing more than a classic miscommunication or misunderstanding.

The second law of thermodynamics asserts that the entropy of the universe increases with time. This is the only law in all of science that asserts a *direction* to the flow of time. This law supports the concept of causality, which ensures that causes precede the results we observe. The cliché example used to describe the second law of thermodynamics and causality is that we never see an egg roll off the counter, break upon striking the floor, and then reassemble itself in its comfy little egg carton condo.

Since the second law of thermodynamics doesn't know that the universe is comprised of two hemiverses, it can't be correct. On the other hand, I think it was a fantastic attempt at trying to verbalize the notion of a moving energy barrier separating the hemiverses.

With the model of hemiverses, we know the entropy of the universe is not only *not* going up, it's not budging. Entropy shifts from one hemiverse to the other, but its value is preserved as the sum of the separate values from each hemiverse. Potential energy shifts from the voidverse and increases the amount of entropy or disorder in our observerse. Think about it: When the observerse first "formed," a hypothetical observer would detect no, or very little, matter and the disorder of the "universe" would be observed to be at its minimum. Over time, the observer would perceive the increase in disorder as it filled with more stuff based on his or her observations. This concept is the motivation behind the second law of thermodynamics but it incorrectly captures the phenomenon, based on my understanding. I would, therefore, recommend that this law be revised to more accurately reflect the nature of the universe or be replaced by a new formulation of the law.

A reformulation of the second law of thermodynamics might go something like this: Over time, energy transfers from the voidverse to the observerse, increasing the perceived level of entropy in the observerse, but entropy is conserved between the hemiverses.

In any event, the emergence of observers depends upon the orderly emergence of heavy elements and the emergence of conditions that are suitable for life itself to take over and continue the "terraforming" of its hemiverse. These conditions won't be replicated again until the pendulum swings back following its return trip from its other extreme energy balance point.

Please refer to Figure 11 below to see a depiction of a logarithmic function. This figure actually provides all of the information we need to understand the concepts of the voidverse and the observerse. Look at the point at which the value of time is unity, or one. The coordinates for this point on this particular chart are (1, 0).

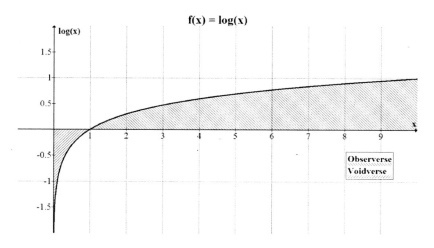

$$f(x) = \log(x)$$

Figure 11. A plot of the function f(x) = log10(x) showing the inverse space-time framework associated with the voidverse (left of x=1) and the space-time framework associated with the observerse (right of x=1).

For this example, let's just forget about the spatial dimensions of our hemi-verses and concentrate solely on the temporal dimension. For any point on the curve possessing a positive exponent, there is a corresponding inverse point that exists on the curve with a negative exponent value. The multiplication product of a value with its inverse value is necessarily (10^0), which corresponds to the value of one or unity. The observerse is always represented by the portion of the curve to the right of unity. On this side of the curve, the area under the curve grows with time as time passes and history accumulates. The value of the function begins to saturate as the barrier pendulum swings from $x = 1$ to its inevitable endpoint in which the observerse achieves its saturated solid energy state.

The voidverse, on the other hand, is always represented by the curve to the left of unity, but larger than zero. On this side of the curve, the value of the function $(\log_{10}x)$ grows progressively smaller while the area between the curve and the x- and y-axes also grows progressively smaller as we proceed to smaller values of x. It's not that time runs "backward" in the voidverse, it's just that the magnitude of time runs "backward," which is to say that it is the inverse of time. I believe the maximum and minimum y-values that this function can attain during the hemiverse transitions are constrained by the combined sum of the energy within the two hemiverses, whereas the lateral distance separating these two corresponding values is proportional to the integrated space-time volume of the universe. In this way, universes may be uniquely defined by the total energy of the composite universe system and the extent represented by its composite space-time; whereas, the physical laws of the universe are defined only by the single quantity that represents its mass-energy density, or put another way, its space-time volumetric mass-energy density.

The observed speed of light in the observerse uniquely defines a universe only at a particular value of volumetric mass-energy density, since this density evolves over time, according to my understanding, as energy flows from the voidverse into the observerse. This mass-energy density sets the minimum curvature of space and, with it, the observed speed of light. The physical laws observed in any and all universes at the point at which the value of the speed of light is identical would likely be observed by its inhabitants to be identical, or at least that's what seems to be consistent with my theory and understanding. Obviously, getting these observers together to share their notes might be a problem, since they are precluded from knowing about any other universes and each other. Besides, ours may be the only universe, anyway.

Time is therefore observed to flow in one and only one direction for the simple reason that observers are around only during the time that flows in the observerse space-time framework with its positive spatial-temporal dimensional exponents. In the "negative exponent" voidverse framework of inverse space-time, there are no observers to perceive any alternate flow of time. Put a simpler way, it seems clear to me from the theory I propose that observers are only present under the same circumstances in which the rules of causality apply or they wouldn't be there to observe in the first place.

For these same reasons, I feel that time travel is possible in only one way and in only one direction, and that is by waiting for time to pass. This particular understanding disappoints me like you wouldn't believe. The bottom line here is that there is no time travel paradox whatsoever. No one can go back and kill his or her grandfather. No one disappears out of photographs when the past gets altered. The past is never the direction through which an observer may pass. An observer may only pass through a continuously flowing "present." That just steams me. The time travel paradox exists, once again, only in our minds and no place else in the universe. Darn.

I think there's a perfectly good physical basis underlying the reason why the universe wants the product to be unity. That brings us back to the way that a photon sees the universe. Recall that the universe appears only as a timeless, spaceless point to a photon scurrying around the universe. This single, non-dimensional point is the dimensionless unity that the universe may very well be. Just as zero represents the sum of precisely identical amounts of "something-ness" and "anti–something-ness," unity represents the product of precisely identical inverse amounts of space-time and inverse space-time. The product can never be anything but unity. Newton's second law of motion, and Einstein's extensions of this concept into relativistic conditions, are only trying to capture this simple underlying relationship, but Newton's and Einstein's equations attempt only to capture or portray events from the perspective of a single hemiverse, namely the observerse.

Anyway, I believe that there's another really interesting aspect to this portrayal of the unfolding or folding of the time dimension that I proposed. The "positive exponent" portion of the curve is strongly related to the "negative exponent" portion of the curve. In fact, even though it's the same darn curve, observers might be inclined to conclude that the dimension of time in the voidverse is fundamentally different, since observation of it is not permitted by the hemiversal energy balance barrier. So, mathematicians might be inclined to portray the voidverse component of the universe with different dimensionality.

The two portions of the time curve represent the two dimensions of perceived time representing the observerse and the voidverse, yet they are two sides of the same coin and really represent the same dimensionality. I'm strongly inclined to apply the same logic to each of the eight seemingly disparate dimensions of space-time comprising both hemiverses, so that proper mathematical representation may enable the number of independent dimensions to be reduced to the four with which we are familiar: x, y, z, and time. The remaining seven dimensions (four of space-time and three representing the bounded surface of the universe system) may then be sufficient to construct an accurate first principles model of a universe, even if the complete set of particular parameters uniquely defining our universe remains elusive and indeterminate.

Furthermore, since the value of the universe is dimensionless unity, the three spatial dimensions needed to represent the universe's outer envelope degenerate down to zero extent. These three degenerate dimensions can then be removed from the total count. Physical phenomena within the universe can then be fully described by only four-dimensional (4-D) space-time.

I told you these last chapters would be short, sweet, and dense, didn't I?

Chapter 35

Implications Involving General Relativity

(The Theory of Gravitation)

Before we start this chapter, I feel the need to vent one more time. Darn, no time travel! I truly hope that nothing in this book deters the screenplay writers of future *Star Trek* movies or series from incorporating time travel in their plots. I'll be so frustrated if time travel plot lines disappear and I'm the one who discouraged sci-fi writers from using the time travel motif!

I suppose that's now in the past, and I should just move on with the present.

If I had a favorite implication regarding my theory, this one has got to be it. I really craved a simpler way than general relativity to understand the relationship between gravity and acceleration. Again, don't get me wrong: Einstein pegged the mathematical description of gravity and its equivalent, acceleration, in his field equations. However, he didn't explain the underlying physical basis for this relationship.

Please recall my proposition that mass is a property of space and not of matter. This represents a huge difference between what Einstein asserted and what I'm proposing.

Einstein asserted that mass is a property of matter, and this particular property of matter causes the curving of space, which he went on to describe mathematically and conceptually in his combined Theory of Relativity and his field equations. On the other hand, I'm proposing that the curving of space results in the appearance of matter in the observerse. Big difference: trust me.

Let's break down my proposition just a bit. The transfer of potential energy from the voidverse takes the form of indeterminate probability waves collapsing down to determinate, and hence observable, matter. Are you with me so far?

So, the shortest possible interpretation of this process is that anything that reduces the potential energy of empty space results in curvature of space and the resulting perception (or appearance) of gravity. Anything, period.

So, Scott, what are the different ways in which you can reduce the potential energy of empty space, you might ask? Easy breezy. There are two very simple ways for any observer to do this, and Einstein captured them. If you're playing the home edition of *The Unobservable Universe*, then observers can do each of these things safely from the comfort of their favorite easy chair.

First, you can perform an observation. The act of performing an observation collapses probability waves constituting the empty space within your observational range into objects and phenomena associated with the collapse of empty space into a puckered state. In fact, this happens whether or not the electromagnetic force is kind enough to send photons your way that you can observe and then construct into a meaningful perception. You can do this just by turning your room lights on and off, or playing peek-a-boo with objects in your vicinity, although the exhilaration associated with the perception of such a commonplace event is pretty much lost by this point in our adult lives.

The second way is just as easy but more poignant to observe. This method involves any sort of acceleration, like flailing your arms about you like you're trying to get someone's attention. The first thing that happens if you do this is that you'll likely get the attention of anyone who sees you. The next step in the process is the really amazing part.

What is acceleration? As we discussed acceleration can be understood simply as the change in velocity with respect to time. If you're stationary (velocity equals zero) or moving at a constant velocity, there's no feeling of either acceleration or gravity resulting from your motion or stillness. Only when you change your velocity will the perception of gravity appear. A common example of the units used in describing acceleration is m/s^2. This means that you are changing your velocity, which is in units of meters/second, over some span of time.

Please recall our discussion involving time flow that the product of obserevse space-time with voidverse space-time must cancel out to dimensionless unity, or a value of one. Well, since there's only one dimension of time in both the observerse and the voidverse, any physical process involving a temporal order of magnitude greater than a value of one does not have a physical dimension with which to compensate. The result is an asymmetry in the local curvature of space with which the phenomenon is associated, which results in the perception of gravity from the direction in which there is a net increasing curvature of space. Don't worry. If you don't get it yet, you soon will after a little more explanation.

Remember Newton: He was the guy who devised humanity's second school of physics after Aristotle and before Einstein. Newton developed three laws of motion that accurately described the circumstances of everyday life on Earth. Einstein extended these laws into the relativistic realm of very high velocities and very large, compact masses. Here's a summary of Newton's Laws:

◊ 1st Law: All objects remain in a state of rest or constant velocity until some outside, asymmetric, external force is applied.

◊ 2nd Law: Force is equal to the multiplicative product of an object's mass and its acceleration, which is represented by the familiar equation, $F = m \times a$.

◊ 3rd Law: For every action there's a reaction that's equal in magnitude but opposite in direction.

Although this is a humongous simplification, Einstein combined Lorentz' transformations with Newton's 2nd Law to produce his Special Theory of Relativity and give the world the famous relationship of $E = mc^2$.

The difference between an object (or observer) at rest and one moving at constant velocity is only the difference in the magnitude of the velocity. In one case, your velocity is zero meters per second; in the other case, your velocity is bigger than zero meters per second. In either case, the units that describe the motion stay the same. In other words, the dimensionality of the phenomenon is the same: meters per second.

As I described, I propose that the universe insists that the multiplication product of space-time and inverse space-time possesses a constant value of unity, or one, and this condition must always be preserved. In the situations involving no motion or constant motion, the product of an observerse phenomenon involving one spatial dimension and one temporal dimension ($f(x,t)$) with the voidverse reaction phenomenon involving one inverse spatial dimension and one inverse temporal dimension ($f(x^{-1},t^{-1})$) possesses the precise value of unity. As a result, there's no transfer of potential energy from the voidverse to the observerse. The potential of empty space surrounding the object is uniform in all directions and no gravity is produced. However, when confronted with acceleration, the universe again does what it takes to preserve a space-time product of unity. It generates gravity to offset the asymmetry to make certain the equation remains balanced precisely at unity.

This explanation provides the underlying reason behind Einstein's assertion that there exists no privileged inertial frames of reference. Any fixed inertial frame of reference is equivalent to any other fixed inertial frame of reference.

Now, let's reintroduce the concept of acceleration and focus on its units.

The units used in our example to describe acceleration are meters per second *squared*. Let's attempt to calculate the product of space-time with inverse space time by performing a simple method frequently used in science called "dimensional analysis." This process is nothing more than the familiar reduction of a fraction down to its lowest common denominator. All we do is make a fraction from the units and cancel out the factors (in this case, units) that appear in both the numerator and the denominator.

Before we do that, it's important to remember that we're limited to the dimensions that are available in the voidverse, namely three inverse spatial dimensions and one inverse temporal dimension. We're limited to the four corresponding dimensions in the observerse, too.

So, in the observerse in which we are performing the observable act of acceleration, we have:
(By the way, the units of gravity are also "meters/seconds².")

$$\frac{meters}{seconds^2}$$

Whereas, in the voidverse, we are limited to a single compensating inverse temporal dimension:

$$\frac{seconds}{meter}$$

Now, onto the dimensional analysis:

$$\frac{meters}{seconds^2} \times \frac{seconds}{meter}$$

leaves us with "seconds" in the denominator after we cancel out the units that are common to the numerator and denominator. There's nothing in our universe that we can do to make this cross product come out to unity. The value of

$$\frac{1}{seconds}$$

can never be made to equal a value of one.

The resulting asymmetry is the phenomenon we refer to as "gravity." That's it. So, when we accelerate through space in the forward direction, we feel gravity tugging at us from behind. If we step on the accelerator pedal in our car, we're propelled forward but we're forced back into our seats. When you accelerate in the forward direction, you actually manufacture gravity behind you and that's why you feel yourself pulled back *into* your seat. This is the physical phenomena that Newton attempted to capture within his three laws of motion. This is the reason why an object at rest resists acceleration.

Okay, so the next logical question to ask is, "How in the world does acceleration through space cause the reduced potential energy level behind us that produces the curvature of space that is the gravity we feel?"

Oh, come on: Give me a hard one! It's easy. As we move through space, let's consider the space in front of us and behind us and forget about the space to our sides that is not relevant to this particular example.

The Atomists noted as they contemplated a fish moving gracefully through the water that there's a difference between the space where we've been and the space into which we are moving. The Atomists, of course, also concluded that motion through space is nothing more than an illusion, but let's just forget this apparently false interpretation once and for all. We haven't yet reached the space in front of us when we're moving, so its state remains pristine and undisturbed, like the driven virgin snow. By contrast, we just passed through the space behind us and you can see our footsteps in the snow. By accelerating through space, we have reduced the number of possible configurations it could take. We've reduced the potential states allowed to it. Put another way, by accelerating through space, we reduce its potential energy. Acceleration provides a means by which we leave telltale "footprints" that reduce the potential energy of the space through which we pass; this, in turn, reduces the number of possible configurations that space can take, and a gravity potential forms.

Take another example. Increase the rate of acceleration and you end up with a bunch of gravity, because the universe provides no way by which you can make the $[(\text{space-time}) \times (\text{space-time})^{-1}]$ product one. Instead, in this case, you end up with units of "seconds^2" remaining in the denominator and you get a crazy amount of gravity. (I will use a shortened version of this Unity Expression throughout the rest of the book: $[(\text{s-t}) \times (\text{s-t})^{-1}]$.)

Take another example. Attach a weight to the end of a string and spin it around a safe distance from other observers and you get a form of gravity we refer to as centripetal or centrifugal force. It's just gravity you're generating from accelerating the weight in a circle through space. The resulting force of gravity is always away from you while the direction of acceleration is toward you. If the string breaks, the weight flies away from you along the tangent represented by its position at the time the string breaks, because the weight makes the transition from steady rotation (angular acceleration) to constant velocity the moment the string breaks. The weight then attempts to take a steady, straight path through space but the curvature (puckering) of space representing the Earth's gravity then forces the weight to take a lower energy path and curve gracefully toward the ground.

When you waved your arms above your head, you alternately accelerated the mass of your arms first in one direction, then in the other direction, and your mind willed the matter comprising your arms to generate gravity. This is what being an observer is all about, and the only place in our universe where you can be an observer is in the observerse.

Let's step this up another notch. You can generate gravity without even going through the schlep of acceleration. You can bypass the acceleration part entirely and generate gravity without expending a single morsel of energy. You can exploit the Casimir effect.

We need to set up a quick thought experiment similar in some ways to others we performed earlier in the book. We're going to walk over to our cosmic vacuum chamber that we had used previously in our cosmic laboratory. We're going to place two very flat, conductive, metal plates in the chamber in such a way that we can move them closer together or farther apart from each other once we pump everything we can out of the vacuum chamber. We're going to place a metal grounding strap between the two plates just to make certain that no electric fields build up in the space between the plates that might affect our measurements. Finally, we're going to attach a pressure measurement device to each plate, so we can detect whether the plates are being forced together or apart. Then, we close the door and turn on the pump to evacuate the chamber.

When the pressure gets as low as our pumps will allow, we begin our experiment. We use our measurement device to confirm that there are no forces pulling the plates together or driving them apart. It reads zero pounds per square inch. Great. Now, let's begin moving the plates toward one another while keeping an eye on the pressure reading from our measurement device. As the plates move closer and closer the meter continues to read zero. The plates are starting to get really close and the meter still reads zero. Then, as the plates continue to approach one another, all of a sudden we see the meter jump to life. We stop moving the plates any closer together when they are only separated by a distance corresponding to tens of atomic diameters. Then, we shift our attention to the pressure meter.

And ... out comes Cuddles; it's time for a pop quiz! No matter which way you answer the following question, you know you won't have to face the indignity of patting Cuddles' rump, so don't break into a sweat.

Here's the question: The pressure meter is showing one of two possibilities. Either the pressure meter is showing a negative value, meaning that the plates are being pulled toward one another, or the meter shows a positive value, corresponding to the plates being pushed apart. Is the meter showing a positive or a negative value?

Think carefully about the situation before answering. What about the position of the plates could possibly cause a force to appear?

Here's the answer: The meter shows a negative value meaning that the plates are being pulled toward one another. This is the Casimir effect.

Oh, I almost forgot. If you answered correctly, then please pat Cuddles on the head in your mind.

Here's what's happening. We created an intensely small gap between the plates. When this gap is small enough, the number of possible configurations that space can take becomes constrained. If we further reduced the spacing between the plates, we would further constrain the variety of possible configurations available to this vanishingly small volume of space. Put another way, we reduced the potential energy of the space within the gap. With the decrease in potential energy, the space puckers and—poof—we manufactured gravity out of thin air. The best part, of course, is that we didn't have to do any work to manufacture gravity other than sucking out every morsel of matter from the chamber. We can spontaneously manufacture gravity just by constraining space and performing a measurement. You can lower the potential energy of space by any method that prevents the mathematical product of space-time and inverse space-time from achieving a value of unity on its own.

So tell me—just how cool is that?

I may still be aggravated about the time travel thing, but the prospect of harvesting gravity really helps take my mind off of it.

I propose that this simple concept underlies Einstein's General Theory of Gravitation and provides the foundation for a more general theory of gravitation which I will propose shortly.

The $[(s\text{-}t) \times (s\text{-}t)^{-1}]$ product is precisely unity in a fixed inertial frame of reference. When the inertial frame of reference is not fixed, in other words when acceleration or deceleration is involved or when a gravitation field is present, then the $[(s\text{-}t) \times (s\text{-}t)^{-1}]$ product introduces another term to ensure that the resulting value remains unity. (This additional term is described in Appendix 1.) The additional term has the effect in the physical world of manufacturing gravity in order to keep the resulting value fixed at unity. So, when the $[(s\text{-}t) \times (s\text{-}t)^{-1}]$ product would not otherwise equal a value of unity, curvature of space, or gravity results in order to balance out the equation.

My theory suggests a more generalized version of Newton's third law of motion. Whereas, Newton's third law of motion can be expressed as

$$\text{action} + \text{reaction} = 0,$$

which holds true within the observerse, the more generalized third law of motion equation can be expressed as

$$[(\text{action in observerse}) \times (\text{reaction in voidverse})] = 1.$$

This generalized third law of motion also preserves the total amount of energy between the hemiverses.

It is possible then that Maxwell's equations, Lorentz' transformations, and Einstein's Combined Theory of Relativity might all collapse down to a unified framework as a result of introducing this generalized third law of motion.

This single law may then underlie all physical phenomena in our universe. If so, this single law becomes the foundation for the Theory of Everything.

Chapter 36

Green Energy

The pursuit of cosmological understanding, as you may have noticed, is quite colorful. It's fraught with terms and references like light, dark, black, blue, red, infrared, and ultraviolet. The assortment of the fundamental particles we call quarks is referred to as "color." The subdiscipline of quantum mechanics that focuses upon the strong force is even called quantum *chromo*dynamics (QCD). Cosmology seems to be simply obsessed with analogies involving color. But "green" isn't one of them. By "green" I'm referring to the thankfully growing trend involving sustainable energy generation or, more accurately, to environmentally responsible power generation and energy storage.

By "environment" I mean Earth's environment. More specifically, I mean preserving an environment on Earth that preserves humanity's continuing ability to live on Earth. If we want to continue holding the discussion on the meaning of life then, first and foremost, humanity needs to ensure a habitable planet capable of sustaining the life that's holding the conversation. It requires that we continue to preserve our observer status. We lose our say in what happens in the universe if we can no longer participate in interacting with it through perception and observation of any sort.

From this perspective, it would seem abundantly clear to me that each political party in America must have a strong interest in preserving a habitable planetary environment, even if they don't exactly share all possible perspectives on the issue. After all, people choose to become politicians and enter the world of politics because they want to have their say in the decisions that get made. If observers lose their say in what happens, then becoming a politician loses its luster, despite the really good benefits and perks that come along with the job. Simply put, if there are no more observers on the planet, no one gets the opportunity to be a stuffed shirt in front of the press and no one gets the great health plan offered to government employees. If there are no observers, there are no health plan administrators, no reporters, no public, no voters of any party affiliation, and no votes by which to enter or remain in elected office.

In any event, this brings us to an interesting point in our discussion. Hopefully, by this point in the book, you understand and agree with my assertion that observers play a pivotal role in forging the state of the universe that we observe. According to my theory, the simple acts of observation and measurement shift the balance of the universe's energy state by forcing indeterminate probability waves into observable, determinate states. At its most basic level, these simple acts result in the flow of quanta of energy from the voidverse to the observerse, thereby creating the particles and gravity that we observe, either directly or indirectly. So, the transfer of energy between the two hemiverses is something that every living entity does all of the time whether or not he/she/it is thinking about it. Our brains are pumps that transfer potential energy from the voidverse to the observerse.

There are times we really do try to think about it and many of these times involve trying to convert the more familiar, everyday forms of potential energy into forms of power we can and do use. A good example of this involves harvesting the potential energy of water by allowing the water to fall from one height to a lower height. This process is the basis for hydroelectric dams. As the water falls to its lower height, potential energy is transformed into kinetic energy. The kinetic energy is harvested in the form of electricity, and the water appears unchanged as it continues its flow to its destination at a lower height than it had before. Aside from the disruption of habitats and spawning behaviors of other observers we call wildlife, the process is generally considered "green."

It seems pretty likely that we could harvest the potential energy of the universe in an analogous fashion, which I'll discuss in only vague ways. This is probably a concept not quite ready for prime time. So, until I or someone else work out the bigger kinks, discussing it prematurely in the wide open spaces of literature is tantamount to displaying a big piece of living meat with a big target sign on it during hunting season. I'm just not voluntarily going there yet. I'll just calmly and quietly wait for my detractors to get distracted by some other shiny object they can criticize before going there.

Observers cause the transfer of energy between the two hemiverses all the time. That's just what observers do, plain and simple. Therefore, it's my firm belief and hope that a carefully orchestrated set of observations and measurements masquerading as a scientific contraption of some sort will be capable of performing the sustainable, controlled conversion of energy from the void hemiverse to the observer hemiverse, providing us with a humongous amount of power to use for something constructive, like perhaps maintaining the habitability of our planet for the future crops of observers. I believe that a properly refined and coordinated choreography of observer actions is capable of gently and sustainably coaxing out from the universe the power we crave. We can then use this power we so crave to drive the economies of nations that have become mindlessly enslaved to the artificial Economy God, which is often portrayed as real but really is nothing more than a fragile scheme by which the common resources of our world are controlled by the few.

Once again, such a contraption would appear, at least at first, as a perpetual motion machine, while in fact the machine would simply facilitate the flow of energy between hemiverses.

I think this might also be a good time to reconsider the definition of "environmentally responsible." Today, our use of energy limits the mess we can make of the universe to one tiny, incrementally non-inconsequential planet of observers. Sure, if we really mess things up, we don't actually destroy the planet; we just make it uninhabitable. We can be quite confident from examination of the fossil record that the demographics of the observer populations have changed drastically over time. Although some accuse me of a strongly parochial view on the matter, I am personally biased in favor of the current arrangement, which involves not only the continuing presence of humanity but its placement at the top of the food chain. Call me a snob, but I like it and I want to see it remain this way. I'm sure my potential (and not yet existing) grandchildren and great-grandchildren would probably agree with my sentiments when the time comes for them to cast their votes.

Tapping into the potential energy of the universe that is stored in the void hemiverse, then, might represent a slippery slope, since it offers the prospect of mucking up the universe's environment. The intentional transfer of energy in one direction from the void hemiverse to the observer hemiverse could hasten the universe to a state with too little potential energy (perceived as crowded space by observers) and too much stuff, sort of like the environment of a hoarder. Such a state could be irritating to observers throughout the universe, especially if it caused their devastation. Ultimately, the power we generate during the transfer of energy from the void hemiverse to the observer hemiverse will require us to do something to preserve a healthful, habitable universe for all observers by keeping the potential-kinetic energy balance at a responsible level, whatever that might end up being, and to the extent that it's even possible.

Observers regularly and routinely shift energy between the hemiverses as a result of their actions and observations. This understanding leaves me with little doubt of the prospect that observers, through the proper schemes, may further alter the energy balance by tapping into the universe's potential energy. If possible, such schemes, and the machines based on them, might appear as inexplicable, perpetual, motion machines under the scrutiny of today's science, since energy would be drawn from the voidverse in ways that cannot be observed directly using electromagnetic-based observational techniques. This ultimate green energy production scheme would undoubtedly involve constructing a machine that exploits the Casimir effect, or a combination of the Casimir effect and electromagnetism.

Chapter 37

Long-Distance Communication

By long-distance communication, I certainly don't mean free, unlimited minutes to anywhere in North America and Europe. I mean really long-distance communication.

Hypothetically of course, just imagine that I represent a member of an intelligent species. I think about how absolutely cool it would be to find another intelligent species in the universe with which I could communicate and continue my dialog about the nature of the universe after I've bored everyone on my home world. What's a guy with a lot to say to do in such a circumstance? Well, I don't actually know whether there's other intelligent life out there, but I considered the statistics and decided that the prospect of other intelligent life is likely, or at least possible. What's next?

I do a quick rundown of the communications options available to me. Strong force? Sure, it's strong but its range is severely limited. Think of a big, strong T. Rex with arms about 10^{-15} m long. That won't help. Weak force? Despite its name, it's pretty darn strong but with an even shorter range. Like a somewhat smaller T. Rex with arms a thousand times shorter than those of the Strong Force T. Rex.

Let's continue. Electromagnetism? Yeah, that's good because it's pretty strong, even stronger than the weak force. Unlike the weak force, its range is unlimited. It can reach clear across the universe but it drops with the square of the distance. The nearest most likely places to find a cross-species discussion partner are still pretty far away and that causes a couple of problems. First, the strength of the signal is going to drop off like mad, so the signal might not even be noticed on the receiving end. Second, it's going to take time to get a response.

In fact, it's going to take a great deal of time and far more time than would allow a practical conversation. It might take years or even millennia to get a response to my message of "I'm sorry, I didn't catch that. Could you please repeat it with a fantastically stronger signal?" Of course, I could boost the strength of the signal by using a focused, collimated, coherent beam of light in the form of optical laser communication instead of radio waves, but then I would have to have a really precise idea of where to point my laser communication apparatus, which I won't have until someone out there communicates with me in such a way that I could detect the signal and determine its origin.

Oh gosh, that leaves me with gravity. Quite frankly, gravity is the equivalent of the kid in grade school that no one wants on their dodge ball team. Beggars can't be choosers, right? So here goes. Gravity? That's just an awesomely bad idea for lots of reasons. I don't even know where to begin. It's twenty-five orders of magnitude weaker than any of the other forces. If I could send out a gravitational wave, and the existence of gravitational waves is far from established, as soon as I send it out, it'll just drop like a very puny lead balloon with the square of the ginormous distance involved. Creating a meaningful gravitational wave in my local vicinity will likely pulverize and vaporize me, anyway, so perhaps I ought to just forget that idea. Yeah, that's right: Just pretend like you never thought about the silly prospect of communicating via gravity.

Well, that was a short discussion as my discussions go. But wait, there's more! What about that "spooky action at a distance" entanglement thing?

Hmmm ... well, it doesn't get weaker with distance, and that's a really good thing. Our observational skills and measurement technology permits us to detect the collapse of a probability wave into a determinate state, and that's also a good thing. Of course, we can't compare the determinate state of a particle to the corresponding determinate state of its partner particle because, quite literally, we have no idea where in the universe the partner particle might be. But we can reasonably suspect that entangled particle partner pairs have been sifting through the universe for billions of years filling every nook and cranny. So, how then might I make use of entanglement to communicate with someone far away? It would be tricky to be sure, but I propose that at least one prospect is possible.

Here's how the transmission side of the communication device would work. My proposition is that entanglement represents a fundamental, common, universal phenomenon in which partner particles have already been conveniently scattered throughout the universe. Sending a signal out through entanglement is as easy as performing a structured set of simultaneous measurements on some fairly large number of photons or fundamental particles. Such a structured set of measurements could correspond to the stream of 0s and 1s constituting a binary data stream representing an encoded message. By performing this structured set of measurements, the transmitter is establishing a structured set of corresponding entangled states throughout the universe. The base frequency of the clock used for streaming out the binary coded entangled states represents the carrier wave upon which the information is transmitted.

The receiving side would appear to me to be a bit more complicated. The receiver would need to perform an exhilarating quantity of quantum measurements, such as observations of photon and particle spin values and look for a non-random signal that emerges coherently in time. Such a scheme would likely require a positively outrageous number of detectors and a correspondingly outrageous amount of computational power, but it's possible. As far as the detectors go, it wouldn't even matter where you place them, because spatial location is completely inconsequential when it comes to entanglement.

The World Wide Web provides a useful distributed computational scheme, and such a scheme is used by the SETI project in its SETI@home application. In this scheme, excess computational power of the home computers of its participants get harvested and applied to the problem of detecting a message in the radio signals received by the Very Large Array near Socorro in my enigmatic state of New Mexico. I suppose it would be a good idea to examine some good candidates for the frequency of the carrier wave, such as the base frequencies of important, natural, at-rest phenomena. I would conjecture that a truly intelligent race would select a useful, practical, and obvious base frequency if this race wanted to improve their prospects of finding a cosmic pen pal.

Don't get me wrong: I'm not saying that my suggestion is easy or terribly practical right now. I'm just saying that it might be the most practical scheme that we have available to us by far, and I think a truly intelligent race wouldn't be needlessly wasting its time and energy on any schemes that involve huge waiting times and vanishingly small signal strengths. It would seem absolutely incongruous to me if we received a call from a race of super-intelligent, nearly omnipotent beings and the call dropped like the usual cellular phone call. That would seem as incongruous as Jean Luc Picard staring incredulously at the view screens on the bridge of Enterprise D as they displayed the contents of Microsoft's infamous Blue Screen of Death. It just seems very wrong, that's all I'm saying.

Communication by entanglement bears a strong resemblance to the "subspace" communicators envisioned by Gene Roddenberry introduced with the original *Star Trek* series. Granted, Gene Roddenberry had to dream up some instant communications scheme or else an entire season of *Star Trek* would have involved nothing more than waiting for a message from Star Fleet to announce some mission or emergency, and *Star Trek* would likely not have achieved the cult status it has enjoyed. Then again, some really nerdy science and engineering types (I'm guilty as charged!) would probably have been drawn to the realism that this represented, and *Star Trek* would have still enjoyed having avid fans, albeit a far smaller fan base and fewer reruns.

Chapter 38

Mass Density versus Matter Density

If you've stuck with the book this far, then it's fair for me to assume that you share some of my rather esoteric interests concerning the universe. Now, I'd like to delve a bit further into the concept of mass density and how this value relates to its cousin property, matter density.

Please recall our earlier discussions about the carbon atom, in which we compared its mass density to its volumetric matter density. The former metric is a commonly accepted material property metric. The latter metric I call the matter density, which I made up for the purpose of providing perspective with regard to our perception of solid matter.

The customary units for mass density are grams per cubic centimeter, g/cm^3. By contrast, the matter density metric I proposed is a ratio between the number of parts of space occupied by matter to the number of parts of space occupied by the emptiness we perceive. This will become clearer as we go through examples below.

For both the mass density and matter density calculations, I determined the volume of the carbon atom by calculating the volume of a sphere using an estimate of the atom's "hard shell" radius. In other words, I used some equivalent "hard" radius of the atom's electron shells, which are somewhat squishy, since there's no absolute firm boundary to anything as we've learned. The electrons in these shells are flitting about like there's no tomorrow and we have no precise idea of their locations. Electrons have an electric charge state of -1, so they interact with other charged bits of matter. As a result, we can only estimate some upper value by which we can reasonably expect to find all of the atom's stuff.

It's important to consider this last statement a little more deeply. We don't know the exact dimensions of particles on the quantum scale because we can't. The Heisenberg Uncertainty Principle also ensures that our knowledge of position and momentum both can't be complete, so since there's no hard position information, it becomes very difficult indeed for us to calculate a "hard" volume. And again, I propose this is the case, since we cannot directly observe the properties of empty space and the indeterminate probability waves it possesses.

Allow me to attempt a reformulation of the Heisenberg Uncertainty Principle that incorporates key elements of my theory: "It's not possible to precisely measure both the position (Class 1 observation) and momentum (Class 2 observation) of a particle without inducing a change in the balance between kinetic energy and potential energy. The act of performing a Class 1 observation of a particle's position changes the potential energy of the space occupied by the particle, and with it, the particle's mass and momentum. Conversely, a Class 2 observation of momentum alters the potential of the space occupied by the particle, and with it the position it occupies and the trajectory it takes. Therefore, the more precisely you determine one characteristic, the less precisely you may determine the other characteristic."

As you might also recall, I described the matter density of a carbon atom as *less than 1 part matter to 500 trillion parts empty space*. I can't actually tell you what the precise volumetric matter density of a carbon atom is because there's no "hard" boundary defining it. So, the best anyone can do is approximate it. Improving measurement technology makes it possible to lower the upper bound we place on the size of particles. The combination of results obtained through improving measurement technology and through improving mathematical models has resulted in the steady lowering of the estimates of the sizes of fundamental particles. As a result, I feel quite comfortable rounding my volumetric matter density for carbon to 1 part matter to 1,000 trillion parts emptiness, which in English is 1 part matter to 1 quadrillion parts of emptiness. This ratio is represented by the value of 10^{-15}.

Again, this ratio is based on a very conservative estimate of the volume of a carbon atom. All this ratio really means is that the "actual" ratio using more precise values of the atom's radius once they're available is expected to be even smaller. So, what the ratio is really saying is that we have no idea what the "real" volumetric matter ratio of a carbon atom really is, but we know that it's definitely no bigger than 10^{-15} and it might be considerably smaller. I'm proposing the prospect that the volumetric matter density of a carbon atom might be reduced further until it reaches the ratio that describes the comparative strength between gravity and the other three forces.

Once again, here are the relative strengths of the four fundamental forces.[19]

◊ Strong force: 10^{38}

◊ Electromagnetic force: 10^{36}

◊ Weak force: 10^{25}

◊ Gravitational force: 1

With this in mind, let's have some fun and again compare the mass density and matter density of a few of my favorite things again. See Figure 12.

	Observed Mass Density (approximate)	Calculated Volumetric Matter Density (ratio of matter volume to empty space volume)
Diamond	3.5 g/cm^3	$< 1 : 10^{15}$
Planet Earth	5.5 g/cm^3	$< 1 : 10^{15}$
Milky Way Galaxy	$1.9e^{-23}$ g/cm^3	$< 1 : 10^{38}$
The Universe	$1.9e^{-26}$ g/cm^3	$< 1 : 10^{41}$

Figure 12. Mass Density versus Matter Density.

By the way, the mass density of the universe is equivalent to each cubic centimeter of space in the universe possessing the mass equivalent of about five hydrogen atoms. So, if we could only know the full extent of the universe, we'd easily be able to calculate the mass equivalent of our observer hemiverse. The problem with this is that the full extent of the universe may not be observable to us.

I decided that the numbers in the chart above were interesting enough to see where they would take me if I performed a few relatively simple calculations.

The calculations led me to the interpretation that the actual edges of the universe are unobservable to us, giving the universe its false sense of mystique. I'll tear away this phony mystique in a later chapter dealing with the implications involving dark energy and the Einstein's cosmological constant.

The matter density metric I proposed above provides us with a way to determine the ratio of potential energy in the voidverse to the "kinetic" energy possessed by the observerse. This provides us with good information by which we can determine the current value of the universe's energy balance.

❁

Chapter 39

Implications Involving

Special Relativity (E = mc^2)

As I've undoubtedly mentioned previously, Einstein published his Special Theory of Relativity in 1905. As experimental evidence mounted to support his theory, the remaining fans of the Luminiferous Aether theory thinned out, quieted down, and moved on to other things. The Special Theory of Relativity placed the final nails into the coffin of the Luminiferous Aether fan club following the negative results of the Michelson-Morley experiment performed seventeen years earlier.

Through the Special Theory of Relativity, Einstein established that the speed of light is perceived by all observers to possess the very same value, regardless of their particular inertial frame of reference. This means that observers could be stationary or moving at a constant velocity with respect to each other or to a phenomenon that they observe.

This theory will be forever remembered by the now legendary yet simple equation, E = mc^2, by which the equivalence of mass and energy is established and by which the amount of energy possessed by matter of a particular mass can be calculated. The special theory of relativity established for once and for all the constancy of the speed of light.

Okay, so let's take our usual step back and recall some of the assertions and predictions made by special relativity.

Space-time is unique for each observer. There's no such thing as a single, fundamental, universal tick-tocking of time, since time flows for each observer according to each observer's clock.

Let's return to the somewhat overused example of comparing the clocks of two observers after one had travelled very quickly through space and the other one just stayed put on Earth. Let's take a pair of twins and say they just celebrated their twentieth birthdays. Let's give them names such that we can refer to them easily and simply. Let's call the twin who stays put on Earth, Gaia, and the twin who travels into space, Cosmo. The next day, the twins say farewell as Cosmo is launched into space on a seventy-year mission to orbit around our solar system at a speed approaching that of light.

Seventy years have passed and Cosmo and Gaia are once again united. Gaia, as we expect, is now ninety years old and has clearly and visibly aged. Gaia spent seventy years waiting for this reunion with her beloved brother. Cosmo, on the other hand, sprightly emerges from his spaceship and doesn't show much sign of aging; in fact, he looks just slightly older than the age at which he started his journey through space.

We understand that since the speed of light needs to be the same for all observers regardless of their inertial frame of reference that time passed at different rates for Gaia and Cosmo during the time Cosmo was zooming around the solar system. Now, let's break this down.

First, let's consider Gaia's situation. Gaia was moving through space at a slow, non-relativistic speed. Since she was traveling slowly, she was travelling according to the description that classical Newtonian physics provides. Her slow speed meant that space was not being compressed and—since the speed of light is the same for all observers—if space was not compressed, time also was not compressed. Time ticked away for Gaia at the rate with which observers on earth are usually accustomed.

Now, let's consider Cosmo's situation. In order for Cosmo to perceive the speed of light at the same value as his sister sitting on Earth, both space and time compressed in the way described by Einstein's relativity. Given Cosmo's speed just below the speed of light, both space and time compressed according to Lorentz' transformations in such a way that the ratio between time and space stayed fixed at the speed of light.

Neither Cosmo nor Gaia perceived any gravity resulting from the speed at which they travelled through space. (For this example we need to neglect the gravity that Gaia perceived from the attraction between her and the Earth, since this does not arise from her motion through space. Cosmo did experience gravity during his acceleration and deceleration through space, but let's assume that these events represent a negligibly small portion of the trip.)

First of all, don't you find this example fascinating no matter how many times you consider it? And although both Lorentz and Einstein faithfully and elegantly described the contraction of space-time that results from Cosmo's high speed, don't you ever wonder why those relationships hold?

Well, here's the physical basis for that relationship that emerges from my theory. In any fixed inertial frame of reference, the [(s-t) × (s-t)$^{-1}$] product must remain precisely unity. The ratio of space-time and inverse space-time must remain precisely at unity for all fixed inertial frames of reference. Therefore, in Gaia's "slow" Newtonian-like inertial reference frame, her larger, less compacted space-time needs to be offset by a subsequently smaller, more compacted and imperceptible inverse space-time in order to preserve the product of unity.

For Cosmo and his fast rate of movement through space, space-time is smaller and compressed in such a way as to preserve the perception of a constant speed of light. Cosmo's smaller, more compacted space-time needs to be offset by a larger, less compacted inverse space-time in order to preserve the unity product. This is the physical basis I propose underlying Lorentz' transformations: plain, simple, and elegant.

In essence, according to my theory, the universe isn't intentionally trying to preserve a constant speed of light as the one fundamental truth of the universe. Instead, the universe is insisting upon a [(s-t) × (s-t)$^{-1}$] product of unity. The means by which we observe this singular, fundamental physical law is through electromagnetic-based observation and measurement schemes that provide us with the perception that the speed of light is constant for all observers regardless of their particular inertial frame of reference. The actual value of the speed of light arises from a different physical basis, that of the volumetric mass-energy density of our observerse.

As a result, the real underlying rule is that the [(s-t) × (s-t)$^{-1}$] product must remain unity and the only way we can indirectly detect it is through a scheme that effectively multiplies this product by the speed of light. The constant speed of light is a consequence of the physical nature of our universe; it doesn't describe the underlying physical nature. Only the [(s-t) × (s-t)$^{-1}$] product does.

Chapter 40

The Biggest Mistake Einstein Didn't Make

(Or, A Brief Description of the Origin and End of Our Universe, or The New Dark Age)

I hope I've succeeded in conveying my thoughts on the apparent absurdity of the notion that our universe came into existence as a result of some sort of big bang at which point time started ticking and our universe began its outward expansion that exceeds the speed of light into some mystical "whatever" that creates lots of new space in the process. In fact, in my opinion, the Big Bang model as usually formulated to account for the creation of our universe may be as far as one can get from the implications resulting from my theory.

Let's begin by reviewing some of the interpretations I presented earlier which arose from my perceptions of the universe and the implications of my theory:

1. The universe is comprised of two hemiverses: the observerse and the voidverse.

2. The appearance or existence of the observerse constitutes only one part of a universe, which is comprised of two hemiverses.

3. The existence and state of the universe requires both hemiverses.

4. The $[(s\text{-}t) \times (s\text{-}t)^{-1}]$ product resulting from the dimensionality of these hemiverses must remain unity.

5. The $[(s\text{-}t) \times (s\text{-}t)^{-1}]$ product value of unity is dimensionless, meaning the universe as a whole exists outside the dimensions of time and space.

6. The concept of a moment at which the universe came into being is undefined, since the concept of space-time applies only within the individual hemiverses and not to the universe as a whole. Only the hemiverses have clocks that are reset.

7. The insistence that the universe as a whole came into existence at a particular moment of time is another example of the human mind's propensity to invoke the concept of infinity.

8. The incomplete and flawed perceptions of our universe force us to build a simplified model of our universe that incorporates the concept of infinity.

9. The dimensionless value of our universe is not infinity; it is unity or "1."

10. The Big Bang Theory takes good scientific measurements and builds a faulty model of an expanding universe.

11. We need a better model to account for the appearance of an expanding universe.

In my opinion, Einstein was right the first time about the cosmological constant that he had chosen to include as a fudge factor in his General Theory of Relativity. Well, sort of; he was only approximately correct. Einstein had added the cosmological constant to balance out his equations so the mass-energy of the universe would provide a steady, flat universe comprised of curved space caused by the gravity of matter's mass.

However, my theory proposes that, in essence, Einstein's attempt to describe the universe could really be reduced down to a description of the observerse at a particular value of the universe's energy balance. Like the second law of thermodynamics, the cosmological constant was an impressive, noble attempt to capture the essence of one hemiverse without the knowledge of the other. Knowledge of both hemiverses is required to make thorough and complete assertions about the nature of the universe, so each of these concepts, the cosmological constant and the second law of thermodynamics, represent attempts at making "universal" rules that hold only in very particular circumstances. They represent the truest statements that could possibly be made about the universe given the absence of knowledge and understanding of a universe comprised of two hemiverses.

So, yes, Einstein was right about his decision and struggle to incorporate the cosmological constant. But, he was also keenly aware of the limitations of the concept it represented. Over time, it appears to me that Einstein's confidence level dropped to the point where he was more willing to abandon the notion than to provide a justification for it that he just wasn't comfortable in sharing.

During the 1920s Edwin Hubble collected lots of measurements of the red shift observed in the electromagnetic spectra of distant celestial objects. Hubble analyzed his data and interpreted it as confirmation of Georges Lemaître's hypothesis that led to the development of the Big Bang Theory. In essence, I propose that the timing of events was just ripe for a theory about the universe exploding into existence and with it the creation of space-time. Hubble's data convinced his peers that the universe continued its expansion based on the evidence he collected during his red-shift measurements. So, in 1931, Einstein acquiesced about the cosmological constant and publicly declared it "the biggest blunder" of his life. And the result of all this was that science got stuck in a seemingly incessant discussion about whether the universe is positively curved, flat, or negatively curved.

Quite frankly, Einstein must have felt deeply relieved after abandoning the cosmological constant. After all, I conjecture that he was keenly aware of the bare fact that he couldn't possibly have provided a thorough, rational explanation for its presence in his equations. But, it also seems evident to me that he knew deep down inside that there was an inexplicable but sound rationale for including it in the first place. Without the knowledge that mass is a property of space instead of matter, Einstein could not have created an internally consistent basis for his cosmological constant. At any moment during the evolution of the universe, the cosmological constant can be determined by calculating the multiplication product of the current universal gravitational "constant" and the current mass density of the observerse. Each of these terms vary over time as the universe evolves and the collapse of potential energy from the voidverse results in the observation of more matter and mass in the observerse.

It appears to me, after all, that Einstein deeply understood that the real meaning of the cosmological constant, in fact, was nothing other than the proportionality constant between the universal gravitational constant and the average mass density of a flat universe. The basis for my assertion is that he felt confident that God didn't throw dice to determine the nature of the universe in which we live and that he profoundly understood the universe was flat, even though he had found no way to defend the assertion. I admire Einstein for going out on a limb and including it, although it's clear he was in a "damned if you do and damned if you don't" situation. He informed the world that the universe is flat but understood that his legacy would hang in the balance if he couldn't provide a firm, proper, and succinct mathematical/scientific basis for his assertion. Any attempt to defend the cosmological constant would have left him vulnerable to criticism by his peers that the lofty founder of the world's third school of physics was making a scientifically unfounded, religious-resembling assertion. This was especially true after scientists had interpreted Hubble's data to mean that the universe was expanding and the notion became embraced by the scientific community.

268 THE UNOBSERVABLE UNIVERSE

So, with that little introduction, let me now explain the implications of my theory in accounting for Hubble's measurements. The bottom line here is that Hubble painstakingly collected great measurements, but used them to develop a flawed interpretation. Whereas, Lemaître was just out to lunch, in my opinion.

For this explanation, we need to shift our focus over to black holes, those unimaginably massive (as in really big mass) yet compact clumps of matter for which so much science fiction owes its due. As I mentioned earlier, I propose a definition for a black hole that's dramatically different than those provided by so many others, including Hawking. I propose that a black hole can be represented by a volume of space-time that possesses a vanishingly small amount of potential energy. As I've explained, the collapse of indeterminate probability waves into matter puckers the empty space formerly occupied by somewhat flat empty space into a new shrunken configuration possessing curved space, and this phenomenon is what we refer to as gravity.

In a black hole, the extent of the curvature is huge. Depending upon its attributes, the black hole can bend space to its limit.

I've alluded to my understanding that the speed of light would be infinite in a universe that was comprised of perfectly flat space possessing a mass-energy density of zero. I've further proposed that the mass-energy density of the current state of our observerse means that space-time has a particular minimum curvature. For light to propagate through space, it must always be climbing out of the gravitational well that this base curvature represents. Because light is always trying to climb uphill, its speed is constrained to the value we observe, which is darn close to 300 million meters per second.

As the curvature of space-time grows steeper, the observed speed of light slows down. This effect was predicted by Einstein and confirmed through experimentation over the last hundred years or so.

If we continue to curve space-time into steeper and steeper configurations, we reach a point at which photons trying to climb out of the gravity well just say "screw it" and fall back in under the strong gravity associated with this very steeply curved volume of space. The energy comprising the photon then gets assimilated into this space-time gravity abyss, never to be seen again in our hemiverse. The information that the photon represented, however, lives on in the universe but becomes more than a trifle difficult to observe in the traditional sense.

When the conditions described in the paragraph above are present, we have a black hole. If a naïve photon zipping around the universe gets too close to the black hole, the photon falls into the black hole's space-time gravity abyss and gets trapped. Bye bye, photon; it was nice knowing you. All you can do in this case is hope that the photon hadn't yet spawned any twin-baby, entangled daughter particles along the way, because the tragedy would just become too much too bear.

What we have, then, is an imaginary sphere surrounding the black hole. On one side of this imaginary sphere, photons and their spawn can romp around in safety and security. However, if a photons strays over the boundary of this sphere, it becomes trapped by the gravity of the black hole and there's nothing more it can do about it. It's lights out for that photon.

This radius of this imaginary sphere is called the black hole's Schwarzschild radius, named after Karl Schwarzschild, a physicist and astronomer who gave Einstein's Theory of Relativity quite a strenuous workout. The Schwarzschild radius has come to be equivalent to the black hole's radius.

Let's talk a little more about the nature of black holes. Let's take the perspective of an observer sitting at a comfy distance from a black hole's Schwarzschild radius and watch as light approaches and then passes through the boundary representing the Schwarzschild radius.

Cue the photon: Enter stage center. A photon of a particular visible wavelength arrives on the scene from over our shoulder and we watch it as it moves past us and approaches the black hole. As the photon gets closer and closer to the black hole, we observe the photon's wavelength increasing as it shifts toward the red end of the spectrum. The photon keeps travelling toward the black hole's Schwarzschild radius, as the photon's wavelength continues increasing and increasing, first to the infrared portion of the electromagnetic spectrum, then to the microwave portion of the spectrum. Then the photon reaches and crosses the Schwarzschild radius. End of story.

You might be thinking, "What? The story was just getting good, but the ending sort of left me just hanging."

Here's why the story ends that way, at least from the perspective of an observer outside of the Schwarzschild radius. At the moment that the observer watches the photon reach the Schwarzschild radius, the photon will appear to the observer to be frozen in time and space. At the interface this Schwarzschild surface represents, information is trying to reach the observer at the speed of light, yet the radius represents the barrier at which the speed of light is insufficient for a photon to get its message back out to the oberserver about what it's doing. In essence, the information from the photon reaching us via an electromagnetically based observation scheme appears to an outside observer to be "running in place." The information is in a limbo state frozen in the balance in such a way that the information is being sucked into the black hole at the very same rate at which it's trying to escape and reach us.

On the other side of this Schwarzschild boundary, an exceedingly uncomfortable observer would see the photon complete its journey and lend its energy to the low potential energy clump of matter thereby lowering the black hole's potential energy incrementally. This clump of matter doesn't exist in the form of atoms for the simple reason that there isn't enough empty space to sustain the structure required by the atom. The amount of empty space has fallen to way below the 500 trillion parts of empty space to one part matter ratio of regular atomic matter.

So, let's return to our attempt to find an alternate explanation of the appearance of evidence provided by Hubble to support Lemaître's silly notion about a Big Bang.

As you know, science has recently thrown us the curve that the mass-energy of the universe appears to be outrageously higher than its earlier estimates. The huge amount of the more recently detected mass-energy is in the form that astronomers call dark matter and dark energy. The latest breakdown of the observed mass-energy of the universe is 4 percent of the good, old-fashioned stuff that we've been observing for some time now, 20 percent of dark matter, and 76 percent dark energy. I'm sure that there are lengthy discussions among scientists about whether to alter these percentages by one or two points for each of the three mass-energy buckets, but this is inconsequential.

Here's another little piece of scientific information to throw your way. Please feel free to research what I tell you further, but I'm just going to cut to the chase here. Every single attempt by scientists to measure and calculate the mass-energy density of the universe results in the same value. The value is 1.0 to the best precision of the measurements used in the calculation. The meaning of this value is that the universe is flat. Despite the fact that we just noticed 96 percent of the mass-energy of our universe, every attempt both before and after this extra large heaping of mass-energy was noticed, the calculation of the universe's mass-energy density has consistently been calculated to possess a value of 1.0. Einstein was right. Period.

So, here's another great paradox to slay. Science claims two completely opposing interpretations of its observations. Given the choices for the curvature of the universe of (a) positive curvature, (b) flat, or (c) negative curvature, all indications point to (b) flat, which incidentally was what Einstein was trying to tell us with his cosmological constant. Yet, Lemaître and Hubble started an "expanding universe" fad that's still playing out today and wowing audiences.

Let's destroy this paradox right now, shall we? Perhaps you even see where I'm going with this.

There's a gob more matter and energy in the "universe" than we gave it credit for just a short while ago. The collapse of empty space into matter curves space and produces gravity that we perceive as mass. A lot more matter means a lot more gravity, which is only curved space. Send photons our way from far enough away and the photons need to climb up out of a fairly shallow but unimaginably long gravitational well. The photon is huffing and puffing spending a really long time on its journey to reach us, all the time being acted upon by the force of gravity. As it attempts to climb up this unimaginably long and seemingly never-ending gravitational ramp, gravity causes the photon's wavelength to lengthen and shift toward the red end of the spectrum. Given a sufficiently long journey and the mass-energy density of our observerse, the photon's wavelength shifts to the infrared and then onward to the microwave portions of the spectrum. If the journey exceeds a certain length, the photon's wavelength will get stretched out even more and shift into the radio portion of the spectrum. Longer still, and the photon's momentum is lost under the mounting influence of the gravity through which it has traveled; thus, the photon loses its momentum entirely or reverses the direction of its journey.

In essence, if the universe possesses a sufficient mass-energy density and the universe is sufficient in the extent of its observerse space-time, then an observer will perceive a universe in which he is smack in the center of an inverted Schwarzschild surface. In effect, the observer's perception of the universe becomes limited by the ability of distant electromagnetic waves to weather the gravitational forces they encounter along their paths. If the distance photons travel to reach us exceeds the Schwarzschild radius bubble we live in, then we'll never, ever observe them—that is, except for the photons that just barely make it to the Schwarzschild boundary.

First, let me explain my last comment then I'll go forward to explain the earlier part of the paragraph above.

The photons that make it to the Schwarzschild boundary will appear frozen and steady to us just like the photon in the thought experiment we conducted earlier in this chapter. Those seemingly paralyzed quanta of electromagnetic energy will appear to observers in the center of the Schwarzschild bubble as very constant and uniform sources of infrared and microwave radiation. Guess what? Astronomers and their space-based observatories have confirmed the existence of exceedingly smooth and uniform Cosmic Infrared and Cosmic Microwave Backgrounds, respectively referred to as CIB and CMB. I submit for your consideration that CIB and CMB are not properties of our universe, per se, but rather properties of our observational perspective at the center of an inverted Schwarzschild surface.

Any electromagnetic radiation originating from further out than the apparent edge of the Schwarzschild boundary will never reach the observer, and the observer will think he lives in a universe that is, oh, say, 13.7 billion years old or so. NASA's determination of the age of the universe based on measurements made by its WMAP spacecraft actually informs us, I believe, of the distance separating us from this Schwarzschild surface. Each observer in the observerse will likely perceive they are in a universe bounded by a radius of about 46 billion light-years.

Okay, so what about the electromagnetic waves that originate from within our hemiverse and from within the Schwarzschild surface—in other words, from a place in space closer to us than the Schwarzschild surface? These electromagnetic waves will be shifted toward the red end of the electromagnetic spectrum. Furthermore, the closer the source of their origin is to the Schwarzschild surface, the more red-shifted these waves will appear to the observer. The closer the source of the origin of the photon is to the observer, the less it will appear red-shifted.

That's what Hubble's instruments were trying to tell him. That's what our space-based instruments are trying to tell us. The content of the hemiverse is changing but it's not going anywhere. Any observer who looks to the sky will see only that portion of the universe within the inverted Schwarzschild surface. And each observer anywhere in the hemiverse will get the same wrong impression that everything radiating light is moving away from him or her.

The bottom line? Luminous matter and dark matter together represent the ordinary matter component of the observerse. Dark energy is simply the gravitational effect of the matter residing beyond the event horizon represented by the Schwarzschild surface of an inverted "black hole" at least from the perspective of performing distant observations. Analysis of the CIB and CMB spectral content provide the basis for calculations that ought to establish the lower bound for the size of the universe extending beyond the Schwarzschild radius. Any observer in the universe will perceive the Schwarzschild radius right around 46 billion light-years from his position at the current conditions of the universe.

Over time, as potential energy continues to flow out of the voidverse and increase the observed matter and mass in the observerse, the apparent size of the universe perceived by an observer will shrink as the greater mass density forces the Schwarzschild radius to collapse inward toward observers. Conversely, such a situation would provide the impression of an expanding universe whose rate of expansion was increasing over time.

Ultimately, as the final bits of potential energy transfer to the increasingly crowded observerse, the Schwarzschild radius collapses to its degenerate value as the observerse becomes the new voidverse. The new voidverse is shrouded from the observerse by the Schwarzschild radius that now encloses it and shrouds it from the view of the new crop of observers that will eventually appear on the scene in the new observerse. At that point, the observerse's newest citizens will likely ponder the very same questions we pose about their own seemingly unlikely presence in a universe that had just the right features for life to emerge.

Our universe appears to consist of a topologically contiguous Schwarzschild surface whose surface area grows and shrinks as the energy of the universe flows from one hemiverse to the other and then back, in an overall arrangement as a universe that exists outside of the dimensions of space and time. Ours is a universe that didn't come into being and won't go out of being because it does exist and its existence does not in any way depend on the notions of time and space upon which we, as humans, insist on constructing in our minds as the necessary foundation for reality based on our daily life experiences.

Done. And, just in time for Thanksgiving dinner.

Summary

I t's time to summarize my thoughts and theories and start winding the book down. I propose that there are three errors and omissions of philosophy and physics that underlie all of the paradoxes involving our universe and prevented the emergence of the Theory of Everything. These three errors and omissions are:

1. Failure to understand the nature and structure of the void since the time of Parmenides 2,500 years ago;

2. The assertion from the Copenhagen Interpretation (circa 1930s) that there is no objective physical reality other than that which can be detected through measurement;

3. Mass is a property of space and not of matter, contrary to the prevailing thoughts of modern science and captured in Einstein's General Theory of Relativity (1915).

Twenty-five hundred years ago, the Atomists posed excellent questions about the universe and succinctly captured the illusory portions of their reality in Zeno's paradoxes against motion. The Atomists placed enormous emphasis on the void as an integral part of the thought-based framework they devised in an attempt to understand their world. The four states of being captured within the *Tetralemma*, along with the contemplation of the void, dominated humankind's early philosophical efforts to comprehend the universe, within both Eastern and Western cultures.

The attempted solutions to the paradoxes that emerged from Eastern and Western philosophies, however, neglected to distinguish between the two classes of observations and measurements (direct and indirect) that philosophers, theologians, and early scientists were taking as they began to inventory and organize their perceptions of the universe.

I propose that direct observations and measurements involve only the electromagnetic force. I refer to these observations as Class 1 observations. Class 1 observations involve the detection and measurement of determinate states via electromagnetic intermediaries, called photons, as probability waves comprising the voidverse collapse down to their observable states of matter and particles in the observerse. Examples of Class 1 observations involve only the detection and measurement of matter and particles exclusively through electromagnetic-based schemes that result in information compatible with the spectral sensitivity of the observer's unique sensory apparatus and which therefore enables a perception to be formed within the observer.

The universe is composed of two hemiverses: the *observerse* comprised of space-time and the *voidverse* comprised of inverse space-time.

The transfer of potential energy from the voidverse, in which it is stored, to the observerse results in a reduction of potential energy in the observerse, which manifests itself as curved space; we call this mass. Observations of mass and changes in the potential energy of empty space involve indirect schemes in which electromagnetic intermediaries are used to probe the curvature of space resulting from the presence of mass or the presence of strong or weak nuclear forces. Class 2 observations use electromagnetic intermediaries to indirectly provide information that is otherwise directly unobservable to observers in the observerse. Examples of Class 2 observations include the speed of light in non-fixed inertial reference frames, the measurement of distantly rotating luminous matter, the universal gravitational constant, mass, momentum, and gravity.

I submit that the failure to distinguish between these two types of observations, Class 1 and Class 2, resulted in the perception of paradoxes regarding the nature of the universe. Unification of the four fundamental forces cannot be achieved without recognizing and distinguishing these two observational classes.

The Aristotelian school of physics, which ultimately emerged from the work begun by the Atomists and continued by Socrates and Plato, failed to make the distinction between these observational classes.

Two thousand years later, Newton introduced to the world its second school of physics embodied within his three laws of motion. Since Newton also failed to distinguish between these two classes of observations, his laws of motion did not take into account both of the universe's hemiverses. His laws of motion attempted to capture the nature of the relationship between observed actions and reactions. Newton incorrectly surmised that the conservation of momentum was the foundation for physical phenomena in the universe.

Had Newton known about the hemiverses, he might have correctly surmised instead that the sum of the universe's potential energy (perceived as void) and kinetic energy (perceived as matter) was the foundational relationship that needed to be preserved and conserved. Had Newton been aware of the two hemiverses, he might have formulated a more general version of the third law, or perhaps even a fourth law of motion, in which the product of action in the observerse is offset by a corresponding reaction in the voidverse, described by:

$$(\text{KE action in observerse}) \times (\text{PE reaction in voidverse}) = 1$$

With this "fourth" law absent, paradoxes emerged. As a result of these unaccountable paradoxes, Einstein formulated his Theory of Relativity, which integrated the work performed earlier by Maxwell, Lorentz, and Poincaré. The theory was split into a special relativity and a general relativity to account for the two different types of inertial reference frames, fixed and non-fixed.

Einstein and other scientists of the twentieth century continually failed to make the distinction between the two classes of observations. As a result, the essence underlying physical phenomena continued to be misunderstood as the urge to comprehend the nature and structure of the void subsided following the publication of relativity theory. Since the potential energy of the universe is stored in an unobservable fashion, its essence was considered only peripherally in the equations of relativity. Without the concept of the hemiverses, twentieth-century scientists had no means by which to understand that the conservation of space-time underlies all physical phenomena in the universe. Because of this misunderstanding, mass was presumed to be a property of matter and not of space.

The culture of science has become too rigid to provide solutions to the paradoxes it creates. A fresh look at today's scientific understanding, the road that led to it, and the paradoxes that emerge from it provided me with a bold, new way to view the universe.

According to my framework and my theory, the third and final mistaken interpretation of science that prevented a unified Theory of Everything from emerging was the faulty underlying premise of the Copenhagen Interpretation of Quantum Mechanics that asserts "there exists no objective physical reality other than that which is revealed through measurement and observation."[20]

This assertion is equivalent in my mind to stating that everything that exists must be observed during any attempt at observation or measurement. I propose this assertion is false and must be corrected. A more proper formulation of the premise underlying quantum mechanics might be something like this: physical reality exists independently of our ability to perceive, detect, or measure it.

Furthermore, we are limited in the breadth and capabilities of our observations by the electromagnetic-based sensory organs we possess and these limitations obscure the distinctions between Class 1 and Class 2 observations.

I submit, therefore, that the attempts thus far to establish the Theory of Everything have failed only as a direct result of the following basic laws of our universe and the resulting constraints imposed upon our observation, which continue to be misunderstood and misinterpreted. These are my Laws of Bio-Quantum Relativity (BQR):

1. Mass is a property of space.

2. The sum of the potential and kinetic energies of our universe is fixed.

3. The flow of energy between the voidverse and observerse is the basis of all of our observations.

I submit that these three assertions by themselves provide the corrections to the flawed and incomplete interpretations of today's science. With the emergence of these three assertions, the Theory of Everything not only becomes possible, it becomes unavoidable.

The relationship underlying all physical phenomena in our universe is:

(KE action in observerse) × (PE reaction in voidverse) = 1

This concept can be represented by the following additional relationships and substitutions:

[action in observerse] × [reaction in voidverse] = 1

[phenomenon in space-time] × [reaction in voidverse] = 1

[phenomenon in space time] × [reaction in (space-time)$^{-1}$] = 1

This is the framework I propose for the Theory of Everything.

As energy transfers from the voidverse to the observerse, the observed mass-energy density of the observerse increases over observerse time. This flow of energy causes the universal gravitational constant to increase over time with increasing observerse mass-energy density. Since the universal gravitational constant varies slowly over time, its value passes through a window that is favorable to the emergence and sustenance of life, as it is now and has been for quite some time. Although I see no way to prove it, it seems clear to me that the emergence of life in the universe is fundamental and inevitable.

Had Newton or Einstein developed the Laws of Bio-Quantum Relativity instead of me, one of these scientist-philosophers would have undoubtedly provided us with the framework for the Theory of Everything. Now, let's get our scientists cracking on some new, exciting stuff to read about and to help us heal our world.

If all goes super-duper well, then maybe, just maybe, we'll start seeing scientist-philosophers and statesmen make a comeback in this century. We're going to need them if my theories are correct.

Concluding Remarks

A wise man once said that everything we needed to know we learned in kindergarten. The flipside of that assertion is that everything we learned after kindergarten might have had questionable value. We grew and matured in a world teeming with paradoxes, and each of us were presented with educations that were, in some form or fashion, formulated or crafted to promote the preservation and continuation of doctrinal thinking.

My quest has proven to me that the universe does not create paradoxes; doctrinal thinking does. Above all of Albert Einstein's countless contributions to humanity was a simple piece of advice, "Imagination is more important than knowledge." He understood what I only recently learned for myself.

I had the unbelievably good fortune to sit in a high school classroom and experience a teacher who was willing to part ways with tradition and doctrine to expose me to my world. Mr. Ferri infused within me the wonderment, enthusiasm, skills, imagination, and drive to establish a goal that would ultimately define my life, my identify, my essence, my purpose. I knew with certainty in a way that couldn't possibly be conveyed in words that I had to follow this path to overcome the paradoxes involving our universe or die trying. Mr. Ferri succeeded in providing a terribly confused teenager with the identity that was to define my perception of who I was, of the identity that would be mine. It was the identity I carried whether or not anyone else knew about it. It was the identity that carried me through every challenge I would face in my life. It was the identity that enabled me to find my sacred life partner, my lovely and loving wife, Janna. And, it was this identity that Janna helped me to finally express.

A teacher inspired me to begin a quest that lasted thirty-four years. His positively whacky teaching style instilled within me a sense of awe about the universe and the drive to achieve a better understanding of it to share with the world. He introduced me to the world of quantum mechanics and the mysteries it was ill-equipped to overcome.

A few months ago, after thirty-four years of learning what I could in a variety of disciplines and performing thought experiment after thought experiment, I decided to begin writing a book to capture the essence of a scheme that I thought could be effectively used to overcome the countless paradoxes that philosophers and scientists have described over the course of the last 2,500 years, and which continue to baffle scientists today, along with the emergence of deep, new mysteries and paradoxes.

My initial purpose for writing my book was to establish the basis for a new conceptual framework by which to approach and overcome paradoxes. Instead, I found myself implementing my own methods rather than simply describing the philosophical-scientific approach they represented. Science and philosophy converged in a way that I would not have thought possible only days earlier. Solutions to long-standing paradoxes began to emerge.

In November 2010, the methods I used appeared to me to have yielded the single fundamental relationship governing all physical phenomena within our universe, and my dream of completing the work begun by Albert Einstein may have been realized.

Each and every step of this thirty-four-year-long path was guided and motivated by the values and love that my teacher, Mr. Ferri, instilled in me. I owe him a great deal of gratitude. Only today, on the day I completed the manuscript for this book, did I learn that Charles M. Ferri passed away sixteen years ago. I never took the opportunity to let him know the role he had played in the unfolding of my life.

If my theories are correct, it's quite possible that there will be a beneficial impact on the lives of many people and the future of our planet. Others will need to judge my theories and confirm the predictions provided by my theories before their accuracy or fallacy can be properly evaluated. My book contains the first description of what science refers to as the "Theory of Everything," which underlies the physical phenomena of our universe. The underlying principle is unexpectedly simple, beautiful, and concise, as science had always presumed that it would be. My theories may be right or they may be wrong, but one thing is sure: The perspective I provide in my book will motivate others to view their world in a fundamentally new way. My intention is to provide to others the gift Mr. Ferri provided to me.

The content of this book may be judged well or poorly by those who read it, but I am secure in the knowledge that I remained true to my convictions and true to my identity. That alone is sufficient for me.

I once worked for a manager who taught me that the key to writing winning engineering proposals was to place the "punch line" up front where the reader can be exposed to it easily and quickly. You see, most engineers have the habit of placing the real essence of the proposal's concept at the end in a misguided attempt to build up anticipation to what is almost inevitably an irritating and disappointing conclusion. He taught me that it's just a poor idea to put the punch line where the reader might never find it, if the act of reading the proposal induces the reader to enter a coma before he reaches the punch line.

I want you to know that I took his advice very seriously and implemented it at every opportunity. You might notice, however, that the punch line for this book is very nearly at the end of the book. There's a very simple reason for this—I had no idea whatsoever just a few days ago of what the ending was going to be.

I was stunned when the application of my framework during the writing of my book yielded a single, simple underlying, self-consistent structure to the universe.

I could not express in words the awe and exhilaration I experienced when the simple relationship possibly underlying the physical phenomena of our universe exposed itself. More time will need to pass before I can even fully embrace and accept the result in a meaningful way.

Although I went back into my manuscript to make the development of my assertions and propositions early in the book more consistent with the later content, the book represents a virtual diary of the framework I endeavored to develop and express, as well as its application in definitively overcoming paradoxes, new and old.

The point is that this book did not turn out to be the project I had initially envisioned. It was supposed to represent a framework that I had hoped to share with others as a new thought tool among the many that have been used by scientists and philosophers. My intention was to extend the work of countless contributors into a single, comprehensive framework that spanned the range of disciplines, which I had felt would together provide the clues that explained the emergence of paradoxes and perhaps point the way to possible resolutions further down the road.

Instead, the beauty and elegance of the universe revealed itself to my mind in colors more vivid than my color-blind eyes could ever hope to see. The image revealed a landscape far simpler and broader than any I thought I could possibly imagine. I found the sheer spectrum of implications nearly overwhelming. I'm going to need a great deal of quality time with my family and friends to establish a personally meaningful path for me to follow, in which I can embrace my new relationship with the world and lovingly preserve and nurture the stunning image of the landscape it's already created in my mind.

The underlying foundation of the universe is simple. The perception of boundaries represents a completely unfounded illusion of the world. The boundaries between my internal universe and the external universe don't exist. There's no physical basis whatsoever for the perception that boundaries exist between and among any of us and the universe in which we live.

The implications of my theory are clear. Each observer has but a single life and a single world upon which to act it out.

If my theory is correct, the understandings and predictions it provides hold the prospect for improving the lives of many people and ensuring the long-term health of the planet that spawned and nurtured us.

The content of the book provides the basis for schemes with which we can safely and constructively tap into the vast potential energy resources of the universe. These same resources are available to other living and perhaps intelligent entities that now appear to me more likely than ever to coexist within the confines of our observerse, itself a world limited in both duration and breadth. If such a cosmic community exists, the understanding that might emerge following the reading of this book may very well provide the basis by which humanity becomes a card-carrying member. Whether our presence in the club is desirable or undesirable depends upon the choices humanity makes, both individually and as a species.

However, like so many other scientific developments which preceded it, the prospect exists for the understanding of the fundamental formula underlying our universe to be used in destructive ways. This is just another unfortunate prospect arising from human nature and the nature of our universe. It's my deepest hope that the content of this book provides a means and a motivation by which humankind abandons the boundaries it so often constructs to isolate, trivialize, and "horrible-ize" the other living inhabitants of our precious home world and to deny access of its limited resources to others they deem less fit or desirable.

The result of my work is simple. It provides each of us with an opportunity to view our universe in a new way, and perhaps it will inspire each of us to bravely break down the perceived but false boundaries that limit the respect we feel for ourselves and demonstrate to others. This may be our next best chance to ensure a viable, hospitable environment in which future generations can frolic and thrive. It may even be our last, best chance to do so.

The fundamental concept underlying our universe is clear. The single, dimensionless value of our universe is unity or "oneness." This concept has always been and will always remain the only god I worship.

Acknowledgments

Just as it takes a village to properly raise a child, it takes a community of dedicated friends and professionals to properly produce a book. I want to acknowledge the assistance and support of all those who participated in one form or fashion to this project.

First and foremost, I want to acknowledge my wife, Janna Mintz, for all of her unconditional love, unwavering support, and seemingly endless patience. Janna performed a wider variety of roles during the writing of this book than she might actually have wanted, but her assistance and support were the primary reasons behind the completion of this project. Janna was my most ardent critic, my most enthusiastic supporter, and my most thorough editor. Her endless contributions were a gift of love and devotion, and I am deeply blessed to have Janna as my best friend and life partner.

I want to thank Dr. Robert Lanza for his book *Biocentrism* (Dallas: Ben Bella, 2009), which provided me with new ways to think about observers, the results of the double slit experiment, and space-time.

I want to extend special gratitude to Mike Mostrom for thoroughly reviewing and commenting on my manuscript. I also want to express my gratitude to M. Charles Fogg, Adrian Chernoff, and Harald Schöne for reviewing and commenting on the manuscript. I fully appreciate the challenges that reviewing this manuscript represented.

I want to thank Aren Horowitz, Steve Halpert, Alan Chodorow, and James Cardinale for their open-mindedness, for their patience, and for the many questions and insights they shared with me.

I want to thank Tauby Mintz, my stepdaughter, and her friend, Celia Robinson, for transcribing my handwritten manuscript into electronic form.

I want to thank Mark Tyson, my son, for his constant encouragement and his boundless enthusiasm.

I want to thank Jared Mintz, my stepson, for his support, discussions, the music he composed for my Web site, and his company at the Roger Waters concert in Phoenix.

I want to thank Catherine J. Rourke (www.editor911.com), my editor and friend, for her extraordinary skills, patience, energy, and perseverance, as well as her attention to detail in preparing this manuscript.

I want to thank Thomas Hill for his contributions to the Foreword and for sharing his knowledge of the Greek language.

I want to thank Mary and Andrew Neighbour of Media Neighbours for their extraordinary efforts in helping me expedite the production of this book.

I want to thank Connie Ferri and Stephanie Ferri Green for sharing photographs, memories, and discussions of Charles M. Ferri.

I want to express my deep appreciation to Wikimedia and Wikipedia for creating a community-based repository of human knowledge. Please consider visiting wikimediafoundation.org to make a donation to ensure that this valuable resource is preserved and further extended.

I want to thank Roger Waters for reminding me about what's important.

I want to thank Stephanie Binch, Pamela Gerloff, and Johanna Vandenberg for their ongoing support and encouragement.

I want to thank Dr. Lawrence Marrich for the chiropractic care that kept my body, and especially my wrists, working during this intense book project.

I want to thank Gaylene Garcia for the massage therapy that kept tension and stress under control, and that helped to keep me together during the project.

I want to additionally thank Hugh Barnaby, Paul Stoll, Alan Chodorow, Andy Keyser, and others for offering to read and comment on my draft manuscript.

I want to thank Bejtush Sylejmani for his encouragement and for getting me away from my writing for the breaks and nachos I sorely needed.

Finally, I want to thank Jessie Van Gogh, Oreo (aka Orca kitty), Maleah, and Mr. Cattuccino: the cats that kept me company after I exhausted the patience of my human companions.

Appendix I

Biographical Timeline of Selected Ancient Philosophers

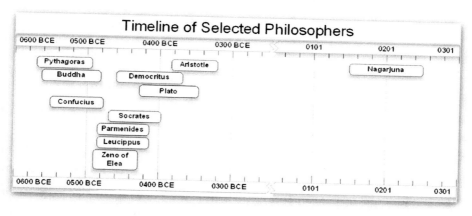

Figure 13. Timeline of Selected Philosophers.

Appendix II

Alphabetical Glossary of
Bio-Quantum Relativity Terms

black hole – A region or volume of space-time within the observerse which possesses little or no potential energy manifesting itself as little or no "empty" space in the observerse; a region or volume of space possessing only or nearly only determinate states

consciousness – the act of synthesizing information about the internal self with information about the external universe

hemiverse – either the observerse or the voidverse; the universe is comprised of two hemiverses, one characterized by space-time and the other characterized by inverse space-time

life – (a) the state of matter which meets all three of the following criteria: (1) possessing self-awareness of internal states, (2) capable of perception, and (3) *incapable* of existing in an *indeterminate* state with an uncollapsed probability field since the state of matter comprising the living entity is under continuous self-observation; (b) the state arising from a self-observable arrangement of matter resulting from self-awareness of internal states

mass – the curvature of space resulting from the transfer and conversion of potential energy from the void (voidverse) into an observable particle in the observer domain (observerse)

matter – determinate states in space potentially observable (measureable) locally by electromagnetism possibly in combination with the weak or strong forces, or potentially observable at a distance directly by electromagnetism in combination with gravity

non-life – inanimate objects and/or phenomena which meet all three of the following criteria: (1) not possessing self-awareness of internal states, (2) incapable of perception, and (3) capable of existing in an indeterminate state with an uncollapsed probability field when not observed

observation – the synthesis of information by a living entity which connects self-awareness with specific information it perceives about the internal self or the external universe

observer – a living entity capable of perception relevant to a specific observation or set of observations; for example, an observation involving electromagnetic energy requires light perception in the correct portion of the spectrum by the observer

observerse – the hemiverse comprised of space-time; the hemiverse portion of the universe comprised of directly observable matter and phenomena arising from the strong, weak, and electromagnetic forces; the mass-energy of the universe not possessed by the voidverse; the hemiverse in which the emergence of observers is permitted and in which causality applies

perception – the collection of information by the self about the internal self (internal states) or about the external universe (external states) through the capabilities of its unique sensory apparatus

Schwarzschild surface – (1) the surface defined by the Schwarzschild radius of a black hole that constitutes the black hole's event horizon; (2) the surface limiting the observations of electromagnetic-based phenomena to the "observable" universe within an "inverted black hole"-like gravity well arising from the mass-energy density of a universe of a sufficient size for the appearance of such a surface

self – a living entity capable of conceptualizing a boundary distinguishing that which it considers internal ("itself") from that which it considers external ("the world" around it)

self-Awareness – the sense of self unique to each self

space-time – the temporal-spatial framework unique to each observer based on the inertial reference state of the observer

universe – (1) the mass-energy comprised of the sum of the observerse mass-energy and the voidverse potential energy; (2) the [(space-time) × (space-time)$^{-1}$] product; (3) the dimensionless unity

void – the potential energy possessed by the void hemiverse which manifests itself as directly unobservable "empty" space within the observer hemiverse; the collective indeterminate states within our universe; the source of the gravitational force; the space-time framework appearing to observers to separate matter into discrete, fundamental particles. Synonyms: vacuum, space, empty space, free space, vacuum fluctuation energy; quantum fluctuation energy; Dirac Sea; Luminiferous Aether; chaos

voidverse – the hemiverse comprised of inverse space-time; the potential energy of the universe directly unobservable to observers in the observerse; the energy of the universe not possessed by the observerse; the hemiverse in which the emergence of observers is not allowed and causality is not required to apply

Appendix III

The Theory of Everything

A more detailed formulation for the underlying Theory of Everything is shown below. A new term, α, is introduced into the unity expression in order to distinguish between fixed and non-fixed inertial frames of reference.

$$[f_1(x,y,z,t)] \times [f_2(x^{-1},y^{-1},z^{-1},t^{-1}) \bullet \alpha] = 1$$

$$[f1(s^m,t^n)] \times [f2(s^\beta,t^\gamma) \bullet \alpha] = 1$$

where:

m, n = any numerical value

$$\beta = 0, -1, -2, -3$$

$$\gamma = 0, -1$$

α = function required to preserve unity value of equation.

A fixed inertial frame of reference exists when $\alpha = 1$.

A non-fixed inertial frame of reference exists when $\alpha \neq 1$.

Gravity is produced under any and all circumstances in which $\alpha \neq 1$.

Appendix IV

Answers to

WordUp! Puzzles in Figure 5

Left puzzle: FEW, SEW, SHE

Middle puzzle: EGO, EGOS, GET, GETS, GIT, GITS, GOT, ITS, MITE, SOT, TIE, ZIG, ZIGS

Right puzzle: APE, APED, APT, ARE, DEAR, DEARS, EAR, EARS, ERA, PAR, PARE, PARED, PEA, PEAR, PEARS, PER, RAP, RAPE, RAPED, RAPS, RAPT, REAP, REAPS, RED, REP, SPA, SPAR, SPARE, SPARED, SPEAR, SPED

Appendix V

Further Reading

Please visit *http://www.theunobservableuniverse.com/links* for a hyperlinked list of additional articles to read on the Web. This list is updated frequently, so please check back from time to time.

Wikipedia offers a broad spectrum of additional content to read on topics covered within this book. Please visit *http://www.wikipedia.org* to access this free content.

Links for further reading about ancient philosophers

Pythagoras – born c. 570 BCE; died c. 495 BCE
   http://en.wikipedia.org/wiki/Pythagoras

Confucius – born 551 BCE; died 479 BCE
   http://en.wikipedia.org/wiki/Confucious

Buddha – (aka Siddhartha Gautama, Gautama Buddha) The time of his birth and death are uncertain; most early twentieth-century historians dated his lifetime as c. 563 CE to 483 BCE, but more recent opinion may be dating his death to between 411 and 400 BCE.
   http://en.wikipedia.org/wiki/Buddha

Leucippus – born in the early 5th century BCE; died in the late 5th century BCE
   http://en.wikipedia.org/wiki/Leucippus

Parmenides – born and died during the 5th century BCE
   http://en.wikipedia.org/wiki/Parmenides

Zeno of Elea – born c. 490 BCE; died c. 430 BCE
   http://en.wikipedia.org/wiki/Zeno_of_Elea

Socrates – born c. 470 BCE; died 399 BCE
   http://en.wikipedia.org/wiki/Socrates

Democritus – born 460 BCE; died 370 BCE
   http://en.wikipedia.org/wiki/Democritus

Plato – born c. 428 BCE; died c. 348 BCE
   http://en.wikipedia.org/wiki/Plato

Aristotle - born 384 BCE; died 322 BCE
   http://en.wikipedia.org/wiki/Aristotle

Books

Bartusiak, Marcia. *Einstein's Unfinished Symphony: Listening to the Sounds of Space-Time.* (New York: Berkley Books, 2000).

Calaprice, Alice. *The Ultimate Quotable Einstein.* (Princeton, N.J.: Princeton U.P., 2010).

Einstein, Albert. *Relativity: The Special and the General Theory.* (New York: PI Press, 2005).

Feynman, Richard. *Surely, You're Joking, Mr. Feynman.* (New York: W.W. Norton, 1985).

Genz, Henning. *Nothingness: The Science of Empty Space.* (Cambridge, Mass.: Perseus, 1999).

Geroch, Robert. *General Relativity from A to B.* (Chicago: U. of Chicago Press, 1978).

Gilder, Louisa. *The Age of Entanglement: When Quantum Physics Was Reborn.* (New York: Vintage, 2009).

Greene, Brian. *The Elegant Universe.* (New York: Vintage, 2000).

Greene, Brian. *The Fabric of the Cosmos.* (New York: Vintage, 2005).

Hawking, Stephen and Leonard Mlodinow. *The Grand Design.* (New York: Bantam, 2010).

Hawking, Stephen. *A Brief History of Time.* (New York: Bantam Dell, 2005).

Hawking, Stephen. *The Theory of Everything: The Origin and Fate of the Universe.* (Delhi, India: Jaico, 2007).

Hogan, James P. *Kicking the Sacred Cow: Questioning the Unquestionable and Thinking the Impermissible.* (Riverdale, NY: Baen, 2004).

Kaku, Michio. *Hyperspace: A Scientific Odyssey through Parallel Universes, Time Warps, and the 10th Dimension.* (New York: Anchor, 1995).

Lanza, Robert. *Biocentrism: How Life and Consciousness are the Keys to Understanding the True Nature of the Universe.* (Dallas: Ben Bella, 2009).

Randall, Lisa. *Warped Passages.* (New York: HarperCollins, 2005).

Rees, Martin. *Just Six Numbers: The Deep Forces that Shape the Universe.* (New York: Basic Books, 2000).

Rosenblum, Bruce and Fred Kuttner. *Quantum Enigma: Physics Encounters Consciousness.* (Oxford: Oxford U.P., 2006).

Siler, Todd. *Breaking the Mind Barrier.* (New York: Touchtone, 1990).

References and Notes

1. Arndt et al., Nature, 14 October 1999.

2. Wikipedia, "Matter Wave," *http://en.wikipedia.org/wiki/Matter_wave*.

3. For the purposes of this book, the author defines "accountable observer" as one who possesses the appropriate sensory capacities and whose presence during and involvement in the experiment is documented or witnessed.

4. Editor's note: The author's statement, "Our interaction with the quantum world altered the external universe …" can be read in different ways. If we come from the understanding that the universe itself exists in objective reality—whether we realize it or not—we can only say that, having come to some greater understanding of the nature of the universe, we as observers (with a rather finite range of observational capacity) have expanded our subjective understanding. Hence, "Our interaction with the quantum world altered the external universe."

5. Wikipedia, "LIGO," *http://en.wikipedia.org/wiki/LIGO*. As quoted on Wikipedia, "Larger physics projects in the United States, such as Fermilab, have traditionally been funded by the U.S. Department of Energy," and some information was obtained from the "LIGO Fact Sheet at NSF."

6. Wikipedia, "Mathematics," *http://en.wikipedia.org/org/wiki/Mathematics*. As quoted on Wikipedia, "The quote is Einstein's answer to the question: 'How can it be that mathematics, being after all a product of human thought which is independent of experience, is so admirably appropriate to the objects of reality?' He, too, is concerned with the Unreasonable Effectiveness of Mathematics in the Natural Sciences."

7. A violin produced by Stradivari or his shop is referred to as a "Stradivarius."

8. Editor's note: This concept was originally derived from the Greek phrase, *he eis to adunaton apagoge*, literally, "the leading away toward that which is not able (to come to existence)." It is found, according to the Internet Encyclopedia of Philosophy, "repeatedly in Aristotle's Prior Analytics."

9. Editor's note: Ontology, the formal philosophical discipline responsible for studying

"existence," is derived from Greek word *ontos*, the participial form of the verb "to be." Participles are verbal adjectives, so they both describe an action and explain its nature.

In terms of what the author is trying to express here, it's more accurate to say that it's impossible to understand the nature of the void without understanding what constitutes "being" and "non-being," which are participial forms of the verb "to be," or "existence," the adjectival form of the verb, "to exist."

Since a verb denotes action—as we know from grammar school—and our understanding of the void is based upon a conception of action words (i.e., verbs). This inherently complicates our understanding of the void, since, the void is unobservable action.

10. Editor's note: According to the dictionary, the word "not" is "used as a function word to make negative a group of words or a word." In the author's explanation, he is using "not" as an adjective to describe the void. Hence, he is actually saying, "The void is not a void," which is clearly not what the author is trying to say. The void is a void—perhaps more so in our understanding than anything else.

Based on what the author is saying there, the Atomists knew that the void existed—which would explain how a fish could swim effortlessly through water. The only thing they could say was that the void existed but it was beyond their ability to understand because of its unobservable state. Logic tells us that it is impossible for non-being "to be." That, in and of itself, is a paradoxical aspect inherent in language—something that the author hopes to bring to the attention of the reader.

11. Wikipedia, "Anthropic Principle," *http://en.wikipedia.org/wiki/Anthropic_principle*.

12. Wikipedia, "Many Worlds Interpretation," *http://en.wikipedia.org/wiki/Many-worlds_interpretation*.

13. Wikipedia, "Multiverse," *http://en.wikipedia.org/wiki/Multiverse*.

14. Robert Lanza and Bob Berman. "Biocentrism: How Life and Consciousness are the Keys to Understanding the True Nature of the Universe." (Dallas: BenBella, 1995), p. 213.

15. Editor's note: The phrase is found in the Fragments, B VI, 2: "... *estin gar einai, meden de ouk estin*," literally, "... for to be is (it exists) as to not be is not (to be)." There are other passages from the same fragments and, while it is not very apparent in this particular sentence, he gets closer to saying that things exist because we think about them.

This is very close to what the author is also saying: Without observers there is nothing to observe and therefore nothing exists (the fallen tree example) without self-conscious "beings."

With respect to the void, we don't have the capacity to observe it—at least with today's technology. But that doesn't mean that we will never have the capacity to observe it. Parmenides, after all, did not have the capacity to observe atoms, such as we do today.

16. Wikipedia, "*William of Ockham,*" *http://en.wikipedia.org/wiki/William_of_Ockham.*

17. Edwin Hubble, "A Relation between Distance and Radial Velocity among Extra-Galactic Nebulae" (1929) Proceedings of the National Academy of Sciences of the United States of America, Volume 15, March 15, 1929: Issue 3, pp. 168–73, communicated January 17, 1929 (Full article, PDF). From *http://en.wikipedia.org/wiki/Hubble%27s_law#cite_note-2.*

18. (New York: Bantam Dell, 1998).

19. Wikipedia, "Four Fundamental Forces," *http://en.wikipedia.org/wiki/Four_fundamental_forces#cite_note-0.*

20. Wikipedia, "EPR Paradox," *http://en.wikipedia.or/wiki/EPR_p.*

Index

About the Author

Visionary physicist, engineer, scientist, researcher and inventor SCOTT M. TYSON has dedicated most of his thirty-year career to probing the far-reaching mysteries of the universe, boldly venturing where few men dare to tread and forever changing how people view the world around them. And he has fifteen patents in space technology and multiple awards to prove it— including a 2011 "Who's Who in Technology" award recognizing him as a key leader of scientific innovations in New Mexico's technology sphere. Tyson spent his high school years on Long Island mesmerized by quantum mechanics, relativity and philosophy, while pretending to be a healthy, normal teenager. After graduating from Johns Hopkins University with an engineering degree, he began his trailblazing career at IBM's VLSI Laboratory, Johns Hopkins University's Applied Physics Laboratory, Sandia National Laboratories, and Westinghouse's Advanced Technology Laboratory.

Responsible for the implementation of new microelectronics approaches, Tyson also served as an advisor to the Office of the Secretary of Defense on space computing technology development and planning, as well as for congressional delegations to accelerate the advancement of meaningful and effective space electronic solutions.

Long-recognized as a pioneering problem-solver and "big picture" futurist in the development of sustainable strategies, Tyson's groundbreaking advances have distinguished him as a change agent in his field. He has published and presented his innovative theories on the future of spacecraft technology at numerous international conferences and professional seminars, establishing meaningful new visions for the space electronics community.

His remarkable ability to integrate seemingly disparate concepts and insights into meaningful and practical solutions and contexts ultimately led to a second career as an author. In his riveting first book, The Unobservable Universe, Tyson deconstructs scientific philosophies to systematically unravel the inconsistencies and assist readers in understanding the puzzling paradoxes of modern science that indicate an incomplete, flawed perception of their world.

Through his thought-provoking yet down-to-earth approach, he builds a compelling new paradigm within a simple, cohesive framework, preparing readers for an inevitable cultural shift that will enable the acceptance of technological change and fundamentally refine the views of laymen and cosmologists alike.

Tyson lives with his wife, Janna, in Albuquerque, New Mexico, where he continues to explore the inner workings of the universe with the same unequivocal passion and fascination of his youth while demystifying it for the rest of humanity.

Photo Credit: Sarah Eastwood, Sarah Love Photography, http://www.sarahlovephotography.com

CPSIA information can be obtained at www.ICGtesting.com
Printed in the USA
LVOW10s2219020114

367878LV00005B/90/P